Reliability Data on Fire Sprinkler Systems

Reliability Data on Fire Sprinkler Systems
Collection, Analysis, Presentation, and Validation

Authored by

Arnstein Fedøy and Ajit Kumar Verma

CRC Press
Taylor & Francis Group
Boca Raton London New York

CRC Press is an imprint of the
Taylor & Francis Group, an **informa** business

CRC Press
Taylor & Francis Group
6000 Broken Sound Parkway NW, Suite 300
Boca Raton, FL 33487-2742

First issued in paperback 2020

ISBN-13: 978-0-367-25185-7 (hbk)
ISBN-13: 978-0-367-77669-5 (pbk)

Library of Congress Cataloging-in-Publication Data

Names: Fedoy, Arnstein, author. | Verma, A. K. (Ajit Kumar) author.
Title: Reliability data on fire sprinkler systems : collection, analysis, presentation, and validation / by Arnstein Fedoy and Ajit Kumar Verma.
Description: Boca Raton, FL : CRC Press/Taylor & Francis Group, 2019. | Includes bibliographical references and index.
Identifiers: LCCN 2019020288 | ISBN 9780367251857 (hardback : acid-free paper) | ISBN 9780429287503 (ebook)
Subjects: LCSH: Fire sprinklers—Reliability. | Fire sprinklers—Testing—Statistical methods.
Classification: LCC TH9336 .F43 2019 | DDC 628.9/252—dc23
LC record available at https://lccn.loc.gov/2019020288

Visit the Taylor & Francis Web site at
www.taylorandfrancis.com

and the CRC Press Web site at
www.crcpress.com

Contents

Foreword

From time to time, a book gives a surprising new perspective on a familiar subject, in this case, sprinkler reliability. The impression of those outside the fire engineering field is that sprinklers are reliable. After reading this work, it is clear that whilst there is much good work on the collection, analysis, and presentation of sprinkler data reliability, there are some basic ruptures in approaches to the field.

I have worked in safety, risk management, economics, planning, and industrial asset management for many years (operations, maintenance and support management, design for performance, and production performance management), so this book touches the core of my work. The consequences of failure in any type of industry, especially the petroleum industry, can be catastrophic if fire hazards are not treated with care and knowledge. Knowing where hazards are and how to reduce them to an acceptable level is critical. One of the most important tools is knowing the risk of assets not functioning as designed. If the risk is known, it can be reduced with proper testing, inspection, and maintenance. Collecting historical data is crucial.

This book explains how to conduct research and suggests the development of methodologies to collect, analyze, and present reliability data in the safety and engineering field. It is of interest to all personnel working with fire, safety, reliability, and engineering, especially those who need to make decisions. There is probably no book like this one.

Professor Tore Markeset Ph.D.
Head of Department of Safety, Economics and
Planning, University of Stavanger

Preface

This book is primarily based on Arnstein Fedøy's Master's thesis, written at the end of a two-year Master's program in Fire Safety at Western Norway University of Applied Sciences from 2016–2018. During this period, he was also the Managing Director of Slokkesystemer AS (Extinguishing Systems Ltd.). He graduated with a Bachelor's degree in Fire and Safety Engineering at Stord/Haugesund University College—Høgskolen Stord/Haugesund in 2005.

Arnstein's interest in automatic extinguishing systems was awakened when he saw Jim Ford, Fire Chief of Scottsdale, AZ, on the Discovery Channel program "In Blaze" in the 1990s. The concept of having a "fire fighter" on duty 24/7 was very appealing, because the results were smaller fires and fewer deaths and injuries, at a very affordable cost. After completing his Bachelor's degree in Fire and Safety Engineering with a thesis on "Comparison of Water Mist and Sprinkler," he had the opportunity to take a study tour to Scottsdale and meet Jim Ford and his colleagues.

Arnstein worked for a sprinkler company and then for a water mist company. A commonality of the two was that they could solve every fire challenge, and either through sprinkler or water mist system. He started to work for himself in 2010 as an entrepreneur, with the business idea that he should provide the customer the best solution, and where there was more than one solution, the customer could choose.

Because of his interest and expertise in extinguishing systems, Arnstein has worked on formulating standards and guidelines and acted as a board member on Norway's only branch association for contractors, manufacturers, and private and public enterprises for all fire engineering disciplines, Brann fagelig Fellesorganisjon (Joint Organization for Fire Protection Norway).

This inspired him to work on a Master's thesis, and reliability was a natural topic choice. The starting idea was to write about why "reliability" differs from one study to another in the same country (for example FM vs. NFPA) and from one country to another (for example US vs. UK). Could this be explained by different standards and regulations in engineering, assembly, and inspection/maintenance over time? It soon became clear that these contribute to reliability, but he did not yet understand why there are so many different terms. After reading many reports on the subject, he understood that some more basic scientific questions were at stake.

This triggered Arnstein's quest to determine how reliability is understood by the data collector and how the data are collected, analyzed, and presented. It also became clear that this work was not only of interest to him and his supervisors; it should be shared with the whole fire community.

Thanks to his supervisor, Professor Ajit Kumar Verma, he was encouraged to publish his work as a book. Professor Verma also agreed to be a co-author. As a teacher and an author in the field of Reliability, Risk & Safety Engineering, his input has been invaluable.

This work taught both authors that the interconnections of fire science, extinguishing systems, and statistical processing are complex. Many books, reports, and articles (including many not mentioned in the book) have expanded their knowledge. That so many have given their time and effort to do qualitative and quantitative investigations of reliability is impressive. They stand on the shoulders of giants.

Acknowledgements

The work of Arnstein Fedøy would not be possible without the support of his family, Aleksander, Katarina, and Anders, and his best friend and wife, Karin.

His friend, PhD Scholar Rune Zahl-Olsen, provided invaluable help in understanding the world of research.

Arnstein Fedøy wishes to thank the National Fire Sprinkler Network (NFSN), especially Steven Mills and Terry McDermott, for their cooperation and their invitation to visit them and speak at their annual general meeting.

He thanks the professors at Worcester Polytechnic Institute, Fire Protection Engineering Department, for their input and their invitation to be a guest speaker on the topic of reliability.

He extends his thanks to the research team at Factory Mutual (FM) Global and Thomas Roche for its help in arranging a meeting.

He thanks the National Fire Protection Association® (NFPA®), especially Marty Ahrens, for its invaluable help finding reports and articles, answering questions, and meeting to discuss findings.

Last, but not least, he thanks the Brannfagelig Fellesorganisjon (Association for Fire Safety Norway/AFSN) for its economic support of his trips abroad. This work would not be the same without their help.

Authors' Biographies

Arnstein Fedøy (M.Sc.) is CEO and Managing Director of Slokkesystemer AS (www.slokkesystemer.as), Kristiansand, Norway, and has specialised himself in extinguishing systems. Apart from his extensive practical experience in the industry, including testing and developing of water mist, he completed theses on "Comparison of Water Mist and Sprinkler" and subsequently on "Collecting, Analysing, and Presenting Reliability Data for Automatic Sprinkler Systems." He has been a part and leader of committees on standards and guidelines, and a board member of Norway's only branch association for all fire engineering disciplines, Association for Fire Safety Norway. Today he leads the work to develop and implement a new certification scheme for all types of extinguishing systems in Norway.

Ajit Kumar Verma is a Professor (Technical Safety) at Western Norway University of Applied Sciences, Haugesund, Norway, and has previously worked at IIT Bombay as a Professor, was an adjunct at the University of Stavanger, and has been a Guest Professor at the Luleå University of Technology, Sweden. His publications include three edited volumes, five monographs, and more than 250 research articles in journals and conferences. He was the EIC of OPSEARCH, and is the EIC of the Journal of Life Cycle Reliability and Safety Engineering and the founder and EIC of the International Journal of Systems Assurance Engineering and Management. He is on the editorial board of various other journals. He is also the Springer series editor of a three-book series.

1

Introduction

1.1 Background

Since the introduction of extinguishing systems, from perforated pipes linked to water tanks or outlets, to the first automatic sprinkler, produced by Henry Parmelee in 1864, to the first "practical automatic sprinkler" in 1874 (TYCO, 2005), the use of automatic sprinkler systems as a tool against fire loss has been popular. In the beginning, this was driven by success stories about systems that saved buildings, allowing business to continue after a relatively short time. But over the years, with the establishment of building codes and as insurance companies began to ask for loss and risk numbers, there was a need for regulations. Sprinklers were installed differently from plumber to plumber, and this was quickly becoming a nightmare. In 1895, a group of men met in Boston to discuss this and to establish national sprinkler rules for the USA. This was the start of the National Fire Protection Association (NFPA), formally founded in 1896 (National Fire Protection Association, 1995).

NFPA began to work with fire from an engineering point of view and to develop codes and standards. It became one of the most important organizations in the world and has now published more than 300 standards (NFPA, 2018a).

With a set of rules and under the watchful eyes of the insurance business, failures and successes could now be tracked. In the years to come, there was increasing interest in measuring sprinklers' performance and improving the standards. This required new test methods, improved technology, and available statistics.

Although there are several types of stationary automatic extinguishing systems (AES), the focus of this book is on automatic fire sprinkler systems because of the amount of available data. There are data over time, from different countries, and for different regulations and standards.

Reports and articles look at many different levels and assess many different scores of reliabilities, using terms such as success, performance, performance effectiveness, operating reliability, operational efficiency, and effectiveness. NFPA (Automatic Sprinkler Performance Tables, 1970 Edition, pp. 35–39) finds sprinklers in the USA are operating at 79.2–98.2% reliability, and it terms this range as *"satisfactory sprinkler performance"* depending on the individual hazard class; overall, the reliability is 96.2% for all types of hazard classes. Other US studies suggest lower performance is achieved. For example, Factory Mutual (FM) says reliability is only 85% (Miller, 1973).

A study in Australia and New Zealand with data from 1886 to 1986 gives 99.46% reliability (Marryatt, Rev. 1988). In the UK, newer studies suggest 93% (Optimal Economics, 2017). The homeland of the main authors (Norway) has several different percentages. SINTEF, now RISE Fire Research, reports: "For all categories of buildings, the average likelihood of the sprinkler operating (i.e. operational reliability) is about 95% and varies between 92–97% (95% confidence interval)" (Bodil Aamnes Mostue og Kristen Opstad ved SINTEF, 2002). This contrasts with a report from *Opplysningskontoret for sprinkleranlegg* (Information Office for Sprinkler Systems), now *Opplysningskontoret for automatiske slokkeanlegg* (Information Office for Automatic Extinguishing Systems), that says "only 8% of the systems meet the minimum requirements of today's regulations" (Opplysningskontoret for automatiske slokkeanlegg, 2003). These apparent contradictions need to be explained.

There are two common approaches to quantify reliability:

1. Component-based (fault tree);
2. System-based (incident data).

Component-based studies use data from an individual component to estimate the effectiveness of a system. System-based studies incorporate failure data and examine effectiveness on a system level. This book is system-based, as this is a tested method to get data on how a system has historically behaved, the book will then compare data across similar system-based studies.

1.2 Description of the Book and Methodology

This book is based on a Master's thesis in Fire Safety, 2017–2018, at Western Norway University of Applied Sciences, Department of Fire Safety and HSE Engineering, titled "Collecting, Analysing, and Presenting Reliability Data for Automatic Sprinkler Systems" (Fedøy, 2018).

It takes a different approach to the subject of extinguishing reliability, in this case, sprinkler systems, than most literature in the area. Unlike other work, it insists that the principles of fire dynamics, extinguishing, and reliability theory must be understood. This is the starting point of the book.

The book is also unique in that it provides a critical review of the relevant literature, based on the desire to seek out the reasons for the diversity in reported reliability levels and the lack of consensus on important concepts.

The authors find it interesting that there are so few critical reviews in the field of engineering.

After the review, the next task was to find out if data on sprinkler reliability are reliable. Accordingly, the book investigates how data are collected, analyzed, and presented in selected studies to determine if this was done using scientifically accepted methods. The critical review is therefore extended to a qualitative document analysis (Jacobsen, 2015) to examine the work of interest. Document analysis is primarily a tool of the social sciences. While an overview is useful to find out what has been written in a particular area, document analysis is a systematic tool to learn more about the subject of interest. It can be used when:

a. It is impossible to get primary data;
b. A researcher wishes to learn how others have interpreted a situation, event, or data; or
c. A researcher wishes to learn what has been done or said.

This book discusses the critical findings of the document analysis.

Finally, the book looks at developing methodologies and proposals for studies with general scientific value.

The reasons for this research are three-fold:

a. To increase knowledge of the reliability of fire sprinkler systems. Fires kill many people every year and cost a lot of money. Any improvement in knowledge and the application of this knowledge will create better fire countermeasures, and improved sprinkler systems would save lives and money.
b. To give a systematic tool for validating *any* study within the field for engineering. It came as surprise to us how little there was on the subject for the engineering disciplines.
c. To give a systematic account of how to collect, analyze, and present reliability data in a scientific way. To the best of the authors' knowledge, no standards, guidelines, or books on this subject are targeted to the engineering disciplines.

1.3 Limitations

While there are many variables that influence the success and reliability of sprinkler systems, the purpose of this book is to establish a general definition of reliability, validate earlier studies, and give examples of to how to conduct scientific studies.

There is little emphasis on physical fires, except for a general introduction to fire theory with a focus on fire growth and the use of different types of sprinkler systems.

1.4 Fire and Sprinkler Extinguishing Theory

What is a fire, how does it behave, and what happens when automatic water-based sprinklers are engaged? This brief introduction gives some basic answers to these questions.

1.4.1 Fire

A fire is an uncontrolled, non-explosion combustion that releases heat, often but not always as visible flames and/or embers, and smoke in the form of odour, gases, and incompletely burned particles. Gases could be CO and CO^2, and incompletely burned particles are often visible, as, for example, soot. The three prerequisites for fire are flammable material, oxygen, and heat. (Store norske leksikon, 2018)

1.4.2 Combustion

Combustion of solid materials is the chemical process whereby heat decomposes flammable material to the point where it releases enough flammable gas and particles to sustain the fire. Combustion can start with a build-up of heat caused by organic material stacked in such a way that heat released in an exothermic reaction cannot be conducted away; the reaction grows beyond this point to start combustion (smouldering fire). At some point, the smouldering will come to the surface and be in direct contact with air (oxygen); at this point, the heat will be sufficient to ignite the decomposed products. Combustion can also be started by heat from any source, such as exposure to open flames. In such instances, both fire and ignition sources are available, and ignition can start at lower temperatures than the transition from smouldering to open fire (Drysdal, 1998).

1.4.3 Room Fire

A room or compartment fire, under well-ventilated conditions, has three characteristic periods illustrated by Figure 1.1.

Compartment fire without sprinklers

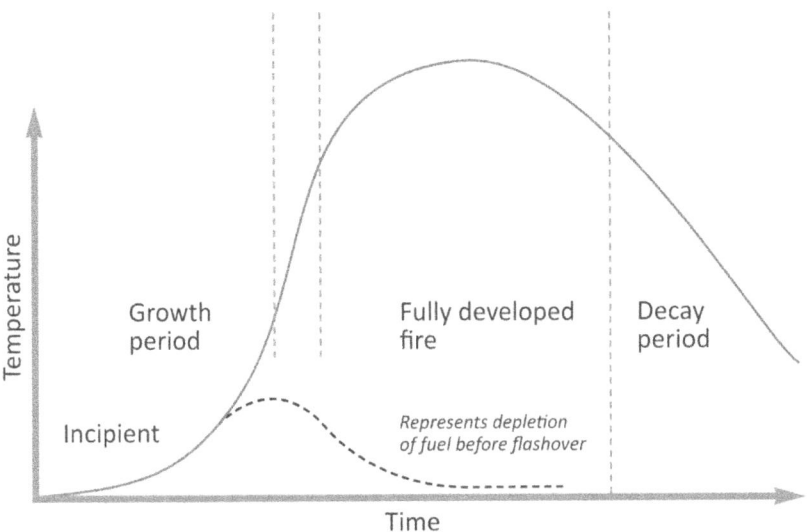

FIGURE 1.1
Classic time/temperature curve for compartment fire.
*Dotted line represents decrease of fuel before flashover.

The first is the growth or pre-flashover period; the temperature is relatively low, and the fire is located close to the origin. As pointed out earlier, a fire can start several ways. Most start slowly, but this does not have to be the case. It can take just a minute (e.g. dry Christmas tree in a living room) or a few minutes: "The mode of burning/combustion may depend more on the physical state and distribution of the fuel, and its environment, than on its chemical nature" (Drysdal, 1998, p. 1).

The fully developed fire period can start with a flashover or with a post-flashover fire. All combustible materials are involved, and flames seem to fill the space.

In the third period, the decay period, the amount of material in the flames is reduced and the temperature/heat release rate drops to 80% of its peak value.

1.4.4 Sprinkler: Control and Extinguishing

Sprinklers have been used since the 1860s because water is cheap, available, environmentally safe, and easy to use as a measure to control or extinguish a fire. In addition, the construction of sprinklers is relatively simple, requiring only valves, an alarm, and sprinkler heads. Water has the capability to absorb heat, to make flammable materials like cellulose products "inflammable" by filling the surface with water, and to vaporize droplets and displace oxygen (inertization). This ability corresponds to what has been called the fire triangle (Society of Fire Protection Engineers, 2016).

Little research considers the effect of different types of sprinklers; there is much more discussion of hydraulic calculations. An example is the SFPE *Handbook of Fire Protection Engineering*, Fifth Edition, where there are no chapters on sprinklers, but there is one on calculations. However, some work looks at residential sprinklers, storage sprinklers, and other specialized types of sprinklers, including test protocols and reports. The most common views of the effect of sprinklers are shown in Figure 1.2. Simply stated, the fire ignites and grows until the sprinkler activates.

Sprinklers control/extinguish compartment fire

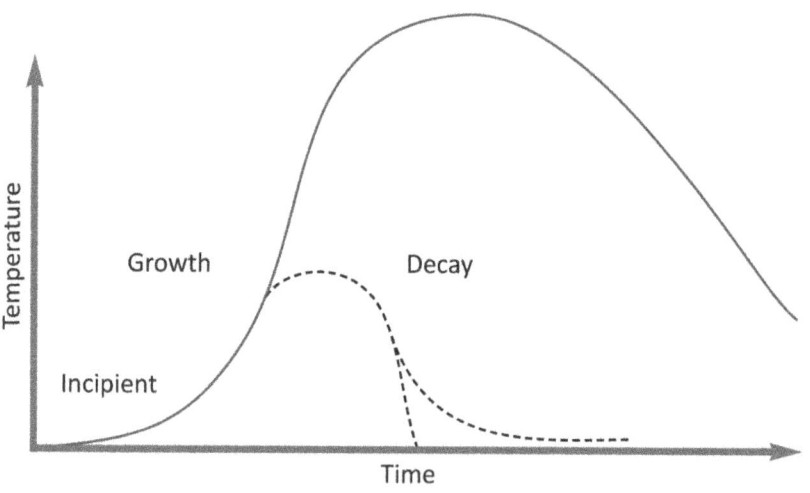

FIGURE 1.2
Time/temperature curves of sprinklers controlling/extinguishing a compartment fire.
*Dotted line represents the effect of sprinkler system in action.

Even if there is little research on the effect of sprinklers as fire control or extinguishing systems within a building, some studies suggest sprinklers extinguish fires in under 1% of all fires (the percentage applies to all fires regardless of whether a sprinkler is present or not for the 1989–1998 period;

see Table 13 in National Fire Protection Association, 2005). The same report estimates that 7.2% of all structures in 1998 had sprinklers (see Table 1 in NFPA, 2005), and in 94% of the cases of fire, the sprinkler operated (see Table 6B in NFPA, 2005), but only 12% of all fires in buildings with sprinklers were extinguished. This make the word "extinguishing system" problematic. Furthermore, the report estimates that sprinklers failed in 7% of structure fires (including operational failures).

Overall, a sprinkler system has four possible outcomes: not operate, extinguish the fire, control it so that manual intervention (e.g. fire brigades) can put the fire completely out, and function so that it postpones flashover to the point where sprinklers have no more control (if there is no manual intervention).

1.4.5 Sprinkler: Postpone Flashover

In recent years, large fires in residential buildings like the Avalon apartment complex in Edgewater, New Jersey (NFPA, 2018b), and other buildings like warehouses (NFPA, 2017) have resulted in large property losses, even though they have all had sprinkler systems. Some of the losses are explained by causes like the water being shut off, the presence of larger and/or more flammable material than the design of the sprinkler system was meant to cover, or lack of maintenance. Except in cases when the system is shut off, there are similarities between how a commercial sprinkler system behaves when it is overrun and how a residential sprinkler system behaves.

Let's consider two scenarios. In scenario one, a fire ignites in a commercial building and grows to the point where the sprinkler activates. For unknown reasons or faults in design, obstacles in water distribution, or other reasons, the fire is not extinguished or controlled. It continues to grow. More sprinklers are activated, but this does not extinguish or control the fire.

In scenario two, a fire occurs on a balcony in a residential building/house, protected by a residential sprinkler system like NFPA 13D or 13R. The fire grows; the heat destroys the window and moves to the living room. The sprinkler activates, and the system operates. At the same time, the fire moves into the attic where there are no sprinklers. The fire goes through the ceiling and activates more sprinklers, but the system has little effect at this stage to control the fire spread.

Both scenarios are represented in Figure 1.3.

As the figure shows, there is a time when the fire is under some sort of control (at least for a time). This period may be characterized by lower temperatures (and probably heat release) that postpone flashover, give better visibility, and reduce the production of hazardous gases and particles (due to the water droplets washing some of the smoke and reducing the hazardous environment), and because the sprinkler system is activated, this alarms the

Sprinklers postpone flashover in compartment fire

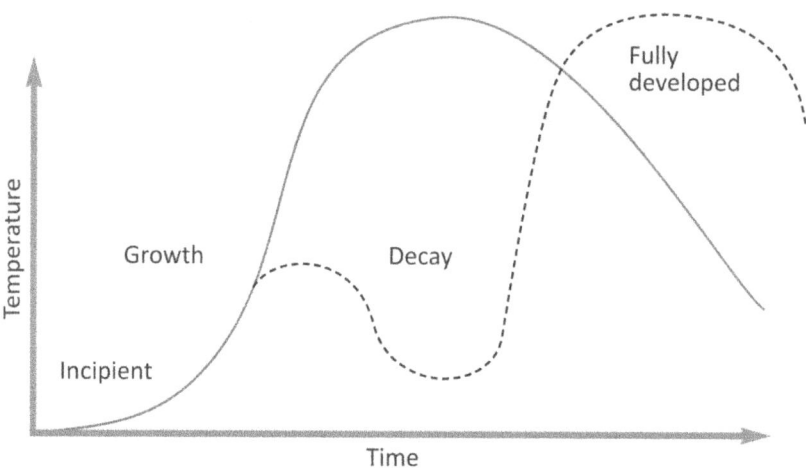

FIGURE 1.3
Time/temperature curves of sprinklers postpone flashover in compartment fire.
*Dotted line represents the effect of sprinkler system in action.

building, neighbourhood, and emergency responders. All these actions help to manage escape and prolong escape time.

Standards on residential sprinkler testing and design indicate strongly that this is the intention for residential sprinklers. UL 1626, "Standard for Residential Sprinklers for Fire-Protection Service," proposes conducting a 10-minute fire test where a maximum allowed temperature (behind the finished surface of the ceiling material directly above the test fire) must not exceed 260°C. This indicates that fire extinguishing is not the aim; rather, the goal is to prolong the onset flashover and increase the escape time. Standards for residential sprinklers, e.g. NFPA 13D, reinforce this understanding:

> 1.2.1 The purpose of this standard shall be to provide a sprinkler system that *aids** in the detection and control of residential fires and thus provides *improved* protection against injury and life loss.
> 1.2.2 A sprinkler system designed and installed in accordance with this standard shall be expected to prevent flashover (total involvement) in the room of fire origin, where sprinklered, and to *improve** the chance for occupants to escape or be evacuated.
> (National Fire Protection Association, 2010a)

The same applied to duration time: "6.1.2 Where stored water is used as the sole source of supply, the minimum quantity shall equal the water demand rate times *10 minutes* unless permitted otherwise by 6.1.3" (*authors' highlight*). According to NFPA, no data support the view that the level of protection goes up when the water lasts longer than 10 minutes for 13D (National Fire Protection Association, 2018, April 11).

1.5 Reliability Theory

What is reliability? The word reliability is often used inaccurately, but here reliability means the ability to function as intended. More precisely, it is the characteristic of or the expression of the ability of a component or system to perform an intended function. This includes the probability distribution of the component or system's lifetime, statistical life expectancy, expected number of failures per unit of time, and the likelihood that something will work at a specific time (Aven, 2006).

1.5.1 Probability Distribution Over the Lifetime

Is a system more reliable when it is new, or after some time? Suppliers and contractors are obviously interested in this question. If a system is very complicated, the chances that they will have to adjust it within the warranty period are higher than if it is simpler. If a system is simple, like many sprinkler systems (except systems with pumps) (Frank, Gravestock, Spearpoint, & Fleischmann, 2013), it is more likely to have an exponential growth in unreliability over time. This is important for the owner, as the cost of maintaining the system's reliability follows the same curve. Probability distributions are used to describe reliability over time.

1.5.2 Statistical Life Expectancy

How long will a sprinkler system or its components last? Perhaps this is not the right question. Perhaps the question is: How long will it take before the cost of maintenance justifies replacing the sprinkler with new components or an entirely new system? Statistical life expectancy is the number of years a component or system is designed to withstand normal load and external stresses.

1.5.3 Expected Number of Failures Per Unit of Time

Since every component has a possibility of failure, every sprinkler system must have a plan for inspection and testing, both by the owner/user and by independent competent persons. Without regular control, the number of failures will increase. This is often measured as mean time to failure (MTF).

1.5.4 The Likelihood That Something Will Work at a Specific Time

When there is a possibility that something will work at any given time, there is also a possibility that it will not work. Owners need to know the likelihood that a device will not fail in a certain time interval or, put otherwise, the likelihood that it will be functional.

1.5.5 Sprinkler Reliability

It is important to separate the strictly operational from efficiency in a sprinkler system. There is a clear separation of the demands placed on a component before it can be used as a part of sprinkler system (testing and approval), especially with the challenges posed by different water supplies and building designs to operate efficiently. All sprinkler standards are written to make sure that an approved component is engineered, installed, inspected, and maintained, including testing, in a way that ensures it will work as intended.

- **Operational reliability** or **operationality** (not to be confused with reliability) is a measure of the probability that a protection system or part of it will operate when needed.
- **Performance reliability** or **efficiency** (performance is probably not the best word; efficiency is likely better) is a measure of the adequacy of the system to successfully perform its intended function under specific fire scenario conditions (fire hazard). In other words, does the system perform effectively in accordance with its design and purpose?

Intended function refers to the correct/proper design following a sprinkler standard. Sprinkler reliability is therefore the ability to function as designed to meet a sprinkler standard with a correct fire hazard.

Sprinkler **reliability** is the ability to function as intended (designed) = **operationality x efficiency** (Barry, 2002).

2

Today's Data

A Study of the Literature

Where should a book like this one start? There are at least three different approaches.

1. Perform a literature search to get a historical overview.
2. Search for the historical development of systems and standards and create an overview.
3. Look for reference material.

This book chooses a mixture of all three, with a focus on studies that stand out because of the number of times they have been cited in other studies, reports, and articles on reliability. Especially helpful in tracking these are literature overviews like *Estimates of the Operational Reliability of Fire Protection Systems* (Bukowski, 1999), *Automatic Sprinkler System Reliability* (Budnick, 2001), and *Reliability of Automatic Sprinkler Systems* (Koffel, 2006). Koffel's overview is presented on the next page. The three above mentioned overviews are used in the review in this book; to compare the studies, see if there has been any development, and emphasize the focus.

Only studies written in English are included in this review; for Norwegian readers, the authors have included data from one Nordic study.

2.1. Relevant Studies

Note that this review does not attempt to judge the value of existing studies. This is not the purpose. Rather, the authors of this book are interested in how reliability data for sprinklers are found.

The overview in Table 2.1. gives the impression that there are many studies of the reliability level of sprinkler systems. However, a short review of the authors and works listed in the table suggests otherwise. Table 2.2. shows how the various authors in the table are referenced over time. Yes and No are used to indicate changes in reference sources over time; the footnotes indicate difficulties this book's authors encountered in their review with source criticism.

TABLE 2.1

Koffel's overview of previous studies

Reference	Reliability of success	Comments
Marryatt[1]	99.5	Inspection, testing, and maintenance exceeded normal expectations and higher pressures. Marryatt's book will be investigated further later.
Maybee[2]	99.4	Inspection, testing, and maintenance exceeded normal expectations.
Powers[3]	98.8	Office buildings only in New York City.
Powers[4]	98.4	Buildings other than office buildings in New York City.
Finucane et al.[5]	96.9–97.9	See comment nr. 6 under Table 2.2.
Milne[6]	96.6/97.6/89.2	From 1959.
NFPA[7]	88.2–98.2	Data provided for individual occupancies; total for all occupancies was 96.2%.
Linder[8]	96	From 1993.
Richardson[9]	96	From 1985.
Miller[10]	95.8	See comment nr. 5 under Table 2.2.
Powers[11]	95.8	Low rise buildings in New York City.
US Navy[12]	95.7	From 1964–1977.
Smith[13]	95	UK data.
Miller[14]	94.8	See comment nr. 5 under Table 2.2.
Budnick[15]	92.2/94.6/97.1	Values are lower in commercial uses (excludes institutional and residential).
Kook[16]	87.6	Limited database.
Ramachandran[17]	87	Increases to 94% if estimated number of fires not reported is included and based upon 33% of fires not reported to fire brigade.
Factory Mutual[18]	86.1	From 1970–1977.
Miller[19]	86	Commercial uses (excludes institutional and residential).
Oregon State Fire Marshal[20]	85.8	From 1970–1978.
Taylor[21]	81.3	Limited database.

[1] Marryatt, H. W. (Rev. 1988). *Fire - A Century of Automatic Sprinkler Protection in Australia and New Zealand - 1886-1986*. North Melbourne, Victoria: Australian Fire Protection Association.

[2] Maybee, W. W. (1988). *Summary of Fire Protection Programs in the U.S. Department of Energy—Calendar Year 1987*. Frederick, MD: U.S. Department of Energy.

[3] Powers, R. W. (1979). *Sprinkler Experience in High-Rise Buildings (1969-1979)*. Boston, MA. Society of Fire Protection Engineers.

[4] Powers, R. W., ibid.

[5] Finucane, M., and Pinckney, D. "Reliability of Fire Protection and Detection Systems," United Kingdom Atomic Energy Authority, University of Edinburgh, Scotland.

[6] Milne, W. D. (1959). "Automatic Sprinkler Protection Record." I W. D. Milne, *Factors in Special Fire Risk Analysis*, Chapter 9, pp. 73–89. Philadelphia, PA: Chilton Company.

[7] NFPA. (1970, July). "Automatic Sprinkler Performance Tables 1970 Edition." *Fire Journal*, 64(4), 5 (35–39).

[8] Linder, K. W. (1993). "Field Probability of Fire Detection Systems," Balanced Design Concepts Workshop. Gaithersburg, MD: National Institute of Standards and Technology Interagency Report.

[9] Richardson, J. K. (1985). *The Reliability of Automatic Sprinkler Systems*. Ottawa, Canada: National Research Council Canada.

[10] Miller, M. J. "Reliability of Fire Protection Systems," Loss Prevention ACEP Technical Manual 8, 1974.

[11] Power, R. W., ibid.

[12] Kelly, K. J. (2003). Trade Ups. Sprinkler Quarterly.

[13] Smith, F. (1982). "How Successful are Sprinklers." *I Fire Prevention, 82*, 28–34. (s. 7). Fire Protection Association 1982.

[14] Miller, M. J., ibid.

[15] Budnick, E. J., ibid.

[16] Kook, K. W. "Exterior Fire Propagation in a High-Rise Building," Master's Thesis, Worcester Polytechnic Institute, Worcester, MA, November 1990.

[17] Ramachandran, G. (1998). *The Economics of Fire Protection*. New York: E & FN Spon.

[18] Kelly, K. J., ibid.

[19] Miller, M. J., ibid.

[20] Kelly, K. J., ibid.

[21] Taylor, K. T. (1990). "Office Building Fires . . . A Case for Automatic Fire Protection." *Fire Journal*, 52–54.

TABLE 2.2
Overview of literature

Reference	Bukowski, 1999[1]	Budnick, 2001[2]	Koffel, 2006[3]
Marryatt (Marryatt, Rev. 1988)	Yes	Yes	Yes
NFPA (National Fire Protection Association, 1970)[4]	Yes	Yes	Yes
Milne (Milne, 1959)	Yes	Yes	Yes
Powers (Powers, 1979)	Yes	Yes	Yes
Factory Mutual (Miller, 1973)[5]	Yes	Yes	Yes? C
Smith (Smith, 1982)	No, B	No, B	Yes
Richardson (Richardson, 1985)	Yes	Yes	Yes
Finucane, M., and Pinckney, D. (Finucane, 1987)[6]	Yes	Yes	Yes
Maybee (Maybee, 1988)	Yes	Yes	Yes
Linder (Linder, 1993)	Yes	Yes	Yes
Taylor (Taylor, 1990)	Yes	Yes	No, B
Kook (Kim, 1990) [7]	Yes	Yes	Yes
Ramachandran (Ramachandran, 1998)	No, B	No, B	Yes

(*Continued*)

TABLE 2.2 *(Continued)*

Reference	Bukowski, 1999[1]	Budnick, 2001[2]	Koffel, 2006[3]
Budnick (Budnick, 2001)	No, A		Yes
Marryatt (Marryatt, Rev. 1988)	Yes	Yes	Yes

[A] New report after comparison was made.

[B] Not included. Reason not known.

[C] There are three references to Miller, Myron J. in Koffel's list.

[1] There are three other listings in Bukowski's "Table 2. Reported Automatic Sprinkler Reliability Data (per cent)" that have *no reference* and are therefore not included here.

[2] There is one listing of US Navy in Budnick's "Table 1." This listing of US Navy has *no reference* and is therefore not included here.

[3] The reference to Kelly's two-page article in Koffel's "Table 1," which has seven studies listed, has *no references*, and is therefore not included here.

[4] The performance in the article is given as *79.2–98.2%*, not *88.2–98.2 %*.

[5] In the reference list, this is listed as: Miller, M. J. (1974), "Reliability of Fire Protection Systems;" Loss Prevention ACEP Technical Manual, 8, 1974. The authors have only been able to find: Miller, Myron J. (1973), "The Reliability of Fire Protection Systems;" at Factory Mutual Research Corporation for The AIChE Loss Prevention Symposium, Philadelphia, PA, November 11–15, 1973, where performance is given as *85%*, not *96%*.

[6] In the reference list, this is given as: Finucane, M., and Pinckney, D. (1988), "Reliability of Fire Protection and Detection Systems," Report Number SRD R431 United Kingdom Atomic Energy Authority Safety and Reliability Directorate, University of Edinburgh, Scotland, p. 15. The authors have only been able to find Finucane, M., and Pinckney, D. (1987), "Reliability of Fire Protection and Detection Systems. Recent Developments in Fire Detection and Suppression Systems" (p. 20). Edinburgh, Scotland: University of Edinburgh, Unit of Fire Safety Engineering, where performance is given as *95%*, not *96.9–97.9%*.

[7] The citation of Won Kook *Kim* is uncertain, because *Kook* is the surname.

In addition to the specific footnote comments in Table 2.2., this book's authors have the following general concerns about one third (29%) of the references:

1. Even if there is a reference to a source, this does not mean the source exists. The reason for this is not known; a source may have been switched, entered incorrectly, or found and then erased/deleted.

2. There is no certainty that cited authors have the right results. As "The Reliability of Fire Protection Systems" (Miller, 1973) clearly indicates, there are reasons to think that a reference can be wrong. Even if there are two articles with the same name, there is no logical reason to think that the expansion of a survey from 1970–1972 to 1973 would rise the reliability from *85%* to *96%*.

3. It is not certain that the authors on the list have read the references. When reading the three overviews and their tables, we

might think that the numbers and references would be corrected over time. This not the case, however, making it logical to think that the work has not been read. This seems to be the case with Kim's "Exterior Fire Propagation in a High-Rise Building," as all the authors have used his first name as his surname. A look at the front page of his Master's thesis would have corrected this. There is also the possibility that they have been read, but the various authors citing him have not corrected the numbers/citations for unknown reasons.

4. References tend to be circular. Since some of the authors on the list have done a comparative study, used former data in their study, etc., they are included by others on their own reference lists. For example, Budnick (2001) is cited by Koffel (2006).

5. If the list is based on references that do not exist, the results are not correct; they only add the appearance of ample data in the area, when this may not be the case.

Since the authors of this book wanted to look at reliability from different countries and at data over time, for their review of the literature, they have selected studies based on four criteria.

The first criterion is that they must be studies of raw data on reliability. Comparative studies or studies using reliability data from other studies, or studies without reference, are excluded. This also means some comprehensive component- and system-based reviews are not included, including the following:

Frank, Gravestock, Spearpoint, & Fleischmann, 2013

Budnick, 2001: Data from others applied.

Ramachandran, 1998: Cost-benefit analysis; data from others applied.

Linder, 1993: Conference presentation; no references.

Finucane, 1987: Conference paper; data from others applied.

Richardson, 1985: Data from others applied.

Smith, 1982: Data from others applied.

The second criterion is applicability; studies done in a small area, like specific building types, or during a very limited time, are excluded, as they have limited applications for this study.

The third criterion is the exclusion of older studies. The rapid changes in sprinkler technology since the end of the 1970s, with the development of quick response sprinklers of different types (like residential and ESFR), must be reflected. Therefore, the following are excluded:

Kim, 1990

Taylor, 1990: The part played by detection and suppression in office building fires; data from 1982–1986.

Maybee, 1988: Only for US Department of Energy, calendar year 1987.

Powers, 1979: Buildings in New York; data from 1969–1979

Miller, 1973: Only for the years 1970, 1971 and 1972; limited data.

Milne, 1959: Old study, pre-1959.

The fourth criterion is the inclusion of relevant newer studies. The National Fire Protection Association has published two major studies since 2006, making NFPA of particular interest, as it is possible to compare the newer work with earlier figures from the same organization. A 2017 study for the United Kingdom (UK) is also included; it fulfils the requirements and is a European study.

Note that only studies written in English are included in the review of the literature. A Norwegian study is included but is commented on separately in Section 2.7.

TABLE 2.3

Overview of relevant studies

Reference	Success, individually and average (%)	Applied /focus/ comments	Comments
Marryatt TA \s "Marryatt" (Marryatt, Rev. 1988)	95.3–100 99.5	Inspection, testing, and maintenance exceeded normal expectations, and higher pressures.	Data from 1886–1986.
NFPA (National Fire Protection Association, 1970)	79.2–98.2 96.2	Data from 1897–1969; 95.8% on average.	Data from 1897; 1924 and 1925; 1969.
NFPA (National Fire Protection Association Research, 2010)	80–94 91	Study done on sprinkler and other automatic fire extinguishing equipment.	Data from NFIRS 2004–2008.
NFPA (National Fire Protection Association Research, 2017b)	81–91 88	Study done only for sprinklers.	Data from NFIRS 2010–2014.
NFSM (Optimal Economics, 2017)[1]	92–97.7 93.6[2]	Study from the United Kingdom.	Data from 2017.

[1] This report uses the terms performance effectiveness and operating reliability. They have been multiplied to determine reliability, as in other studies.

[2] In Table 20, these findings are examined; the percentage given here is *87%*.

There are only two early studies on the list of studies of raw data. The New Zealand (Department of Building and Housing, 2005) explains:

> [We] recognise that there is as yet inadequate data for fire engineering to achieve the accuracy that is expected from, for example, structural engineering. In particular, the probabilities used for a fire analysis must be based on fire statistics derived from a comparatively small data pool of mainly overseas buildings of unknown design. That applies not only to fire scenarios but also to the proper functioning of critical systems including the sprinklers, . . . There appears to be no certainty as to the extent to which those statistics and probabilities are appropriate for use in the New Zealand context.

What conclusions can we make at this point? First, there are irregularities in the references and quoted reliability levels, and second, the list of relevant raw data studies is a short one.

To sum up, this book's authors have the following concerns:

a. The diversity in terms and reliability levels requires explanation;

b. Twenty-nine per cent of the references on the list (Table 2.3) are misquoted and references cannot be found;

c. Incorrect numbers, reference sources, and authors' names go back to the 1970s; and

d. Authors use circular referencing.

Taken together, these strongly suggest that not everything has been done correctly, and therefore a critical review is used on the following chapters.

2.2. Australia and New Zealand's Experiences With Sprinklers

A book that is frequently given credit for showing the success of sprinkler systems is the comprehensive work by Henry William (Harry) Marryatt, *Fire—A Century of Automatic Sprinkler Protection in Australia and New Zealand—1886–1986* (Marryatt, Rev. 1988). This 478-page book takes a close look at the history of sprinklers in Australia and New Zealand, including technical aspects of discharging water and water damage, performance analysis (both general and detailed), safeguarding life, causes of fire, incendiarism,[1] the operation of sprinklers on flammable liquids, electrical equipment, explosions, fires involving high piled storage, smoke and heat venting in relation to sprinklers, fires with large numbers of sprinklers operated, exposure fires, partial protection, fires not controlled by sprinklers, fires

---

[1] The act or practice of an arsonist; malicious burning or inflammatory behaviour; agitation.

involving multiple-jet controls,[2] and economic considerations and cost-benefit analysis.

This book is now in its second edition. It comes with definitions, abbreviations, conversion rates, and explanatory notes.

2.2.1. Reliability

The introduction to Marryatt's book says, *"The record which is detailed in this book show that except under the most extraordinary conditions, it is possible to control fire automatically with a minimum loss of life and property. The evidence suggests that probably there can no better way of safeguarding life and property in the majority of buildings than by equipping them with automatic sprinkler system."* Since the first edition, the term satisfactory performance has been changed to "fires extinguished and/or controlled."

According to the book's definition, *"Fires controlled = Fires which have either been completely extinguished or controlled by automatic sprinkler system to the point that they would be extinguished even if the supplementary action had not been taken by fire brigades or others."*

The book covers 9 022 fires in 231 occupancies, of which 99.46% were controlled (see Preface).

Marryatt's Chapter 4, "Overall Performance Analysis," gives a detailed analysis of the number of sprinklers activated, from one to 113 sprinklers operating and controlling fires (see p. 90, Table 3). This adds up to 8 973 fires and gives a performance/controlling rate of 99.46%. The number not controlled is 49, or 0.54%. The book says:

> The following Table 3 gives the numbers of sprinkler heads operating on all fires except the 49 fires not controlled. Multiple jet controls and spray controls have been counted as each being equivalent to one sprinkler head and these have not been shown separately as full details of the fires in which this specialized equipment was involved are given in Chapter 19.
>
> (see Marryatt, Rev. 1988, p. 90)

On the next page and in the next table, the authors of the present work compare Marryatt's numbers to US numbers. They do not use the NFPA numbers from the 1925–1964 period from this table, preferring instead the Automatic Sprinkler Performance Tables (National Fire Protection Association, 1970); these numbers are from 1925–1969, making them somewhat closer for comparative purposes.

---

[2] Multiple jet or spray controls are thermic controlled valves, often bulb-equipped, that release water from more than one open jet, spray, or sprinkler head over a designed area.

TABLE 2.4
Number of sprinklers operating in US and Australia/New Zealand by per cent and total numbers

Number of sprinklers operating	United States[1]				Australia and New Zealand		
	Wet system per cent	Dry system per cent	Total numbers of fires	Total system per cent	Number of fires	Total numbers of fires	Total system per cent
1	42.6%	20.1%	29 733	37.4%	5 816	5 816	64.55%
2 or fewer	61.0%	32.7%	43 396	54.6%	1 431	7 247	80.41%
3 or fewer	70.2%	41.5%	50 769	63.8%	553	7 800	86.54%
4 or fewer	76.2%	48.7%	55 795	70.1%	290	8 090	89.79%
5 or fewer	80.2%	53.7%	59 156	73.4%	189	8 279	91.84%
6 or fewer	83.2%	57.8%	61 814	77.7%	144	8 423	93.44%**
7 or fewer	85.2%	61.3%	63 724	80.1%	87	8 510	94.40%
8 or fewer	87.0%	64.2%	65 348	82.2%	76	8 586	95.24%
9 or fewer	88.3%	66.4%	66 571	83.7%	50	8 636	95.79%
10 or fewer	89.4%	68.5%	67 629	85.0%	47	8 683	96.31%**
11 or fewer	90.4%	70.3%	68 533	86.2%	22	8 705	96.55%
12 or fewer	91.2%	72.4%	69 464	87.3%	24	8 729	96.82%
13 or fewer	91.7%	73.8%	69 990	88.0%	31	8 760	97.16%*
14 or fewer	92.6%	75.3%	70 788	89.0%	32	8 792	97.51%
15 or fewer	93.1%	76.2%	71 313	89.7%	22	8 814	97.75%
20 or fewer	95.0%	81.0%	73 347	92.2%	59	8 873	98.39%
25 or fewer	96.0%	84.3%	74 464	93.6%	36	8 909	98.79%
30 or fewer	96.9%	86.7%	75 411	94.8%	23	8 932	99.05%
35 or fewer	97.3%	88.6%	75 976	95.5%	12	8 944	99.17%
40 or fewer	97.7%	90.0%	76 472	96.2%	8	8 952	99.25%
50 or fewer	98.1%	91.9%	77 079	96.9%	6	8 958	99.31%
75 or fewer	98.9%	94.7%	77 995	98.1%	10	8 968	99.41%
100 or fewer	99.4%	96.3%	78 533	98.7%	4	8 972	99.45%
200 or fewer	99.8%	99.7%	79 384	99.8%	1	8 973	99.46%**
All fires	100.0%	100.0%	79 544	100.0%	49	9 022	100.00%

[1] These numbers have been updated to the 1925–1969 numbers from the Automatic Sprinkler Performance Tables (National Fire Protection Association, 1970) and are not the original version from 1925 to 1964 shown in Table 2.4.
* This is specified as 96.16 in the original table. This cannot be true and has been changed.
** See point 1 below.

Based on the findings in the table, the authors of this book point out the following problems:

1. Before the table on page 91, Marryatt says: "It will be seen from Table 4 which follows that one sprinkler head in operation was required in 64.55% of fires, while 6 sprinkler heads or fewer were required in

93.36% of fires, and 10 sprinkler heads or fewer were required in
96.16% of fires" (**authors' highlight**). The problem is that the table
gives **93.44%** and **96.31%** respectively (this number has been checked
against a previous table in the book that gives the precise per cent
against number of sprinkler heads operating). Both text and table
cannot be correct.

2. Marryatt's Chapter 15, "Fires in Which Large Numbers of Sprinkler
   Heads Operated," takes a closer look at fires with a large number of
   sprinkler heads, more specifically, fires with more than 10 sprinkler
   heads operating. Marryatt comments: "It has already been shown in
   Chapter 4 that the percentage of fires in Australia and New Zealand
   controlled with six (6) or less sprinkler heads in operation, and ten
   (10) or less respectively, were as follows" (see Table 2.5.):

TABLE 2.5
Fires in which six or fewer and 10 or fewer sprinklers were in operation

Six or fewer sprinklers in operation					
1886–1968		**1968–1986**		**100 Years**	
94%	83 no of fires	91.69%	93 no of fires	**93.39%**	*176 no. of fires*
10 or fewer sprinklers in operation					
1886–1968		**1968–1986**		**100 Years**	
96.7%	85 no of fires	95.13%	96 no of fires	**96.60%**	*181 no. of fires*

However, this is the third place where the authors of the present
work see a difference in the number of sprinkler heads in operation.
If this includes "not controlled," both numbers should be higher
than 93.44% and 96.31%.

3. While the NFPA table separates wet and dry systems, Marryatt's
   book does not. Under "Dry Pipe and Marine Automatic Sprinkler
   Systems" in the same chapter, Marryatt says, "One of the important differences in results when comparing the number of sprinkler heads in operation on fires in Australia and New Zealand
   with those in the United States is that there are very few dry
   pipe systems in these two countries, so few that *no fires* have
   been recorded for this type of installation. However, four fires in
   marine automatic sprinkler systems have been recorded" (*authors'
   highlight*).

4. One activation in the "200 or fewer" row is 113 sprinklers (according
   to Table 3 in Marryatt), giving an overall performance rate of 99.46

≈ 99.5%. This indicates that more than 200 sprinklers operated in 49 not controlled fires. This agrees with Table 4 in Marryatt where the last line, "All fires," puts the number of fires at 49 and the per cent at 0.54. But this cannot be the case. In his Chapter 15, "Fires in Which Large Numbers of Sprinklers Heads Operated", Marryatt "analyses the fires in which more than ten (10) sprinkler heads operated, other than those in which exposure was involved, and where fires were not controlled." This is shown in Table 2.6.

TABLE 2.6
Comparison of the numbers from Table 4 and Table 50 in Marryatt

Number of sprinklers operated	Table 4	Table 50	Difference
11	22	20	2
12	24	22	2
13	31	29	2
14	32	29	3
15	22	19	3
16–20	59	54	5
21–25	36	38	–2

5. Table 50 in Marryatt's book also lists cases where 124, 126, 140, 150, 156, 179, 220, 240, 256, 278, 290, and 361 sprinkler heads were activated. These are not included in Marryatt's Table 4, however. Does this mean Table 10 in Marryatt's book includes exposure sprinkler systems and multiple jet controls and spray controls?

TABLE 2.7
Summary of tables in Marryatt

Number of sprinklers	Table 4	Table 50 in the book	Multiple jet[*]	Exposure sprinkler[**]	Total number (Table 50 + MJ+ES)
11	22	20	1	1	22
12	24	22	1	1	24
13	31	29		1	30
14	32	29	1	2	32
15	22	19	1	2	22
16–20	59	54	2	3	59
21–25	36	38		2	40

[*] Tables 79 to 81 in Marryatt.
[**] Table 56 in Marryatt.

This does not add up, even if it is close for some numbers. When 21 to 25 sprinklers were operated, the gap between all sprinklers, including exposure sprinkler systems and multiple jet controls and spray controls, increases when adding those systems to normal sprinkler systems (if the present authors understand the book correctly).

6. There is no explanation why the 113 sprinklers were considered "fires controlled." For example, the operating area for 113 sprinkler heads is around 113 x 9m^2 = 1 017m^2. This is far more than even the biggest design area (360m^2) for sprinkler systems according to Australian standards. This could perhaps be an exposure system, but this is not known.

7. Marryatt's book has one chapter on partial protection, but there is no evidence of how this is incorporated into performance analysis/tables.

This raises a question on methodology. This is not clarified in Marryatt's book, but there is a clue to this. On page 14 of the Introduction, Marryatt writes: "289 fires were identified from Fire Brigade records in which automatic sprinkler systems operated satisfactorily, but for which no detailed reports were available." And further down the page, he says: "This edition has been dedicated to Wormald International Limited,[3] one of world's largest organizations in the field of Fire Protection and Security and a Company which had the foresight to keep the records which have been so important for so many years."

More information is not found until Chapter 21, "Summary," where Marryatt writes: "Regrettably, this claim could not be sustained for 100 years, because of declining interest in making detailed reports available, the Wormald International Group of Companies being *the only* organization which continued to submit reports to the end of 1986" (*authors' highlight*). It therefore appears that the main source of Marryatt's reports is the largest sprinkler company in Australia and New Zealand. There is no information on scientific independence or how results are tested. The company clearly has a self-interest in a good report.

Because all systems investigated in Australia and New Zealand were wet systems, the present authors now compare Marryatt's numbers with NFPA reports, first looking at the per cent difference between total number of systems (both wet and dry) and only wet sprinkler systems given in the NFPA 1970 report (National Fire Protection Association, 1970), and comparing NFPA's results to Marryatt's. The findings are displayed in Table 2.8.

The average difference between Marryatt and NFPA for only wet systems is 9.7%. The average difference between total per cent and per cent for only wet sprinkler systems is 4.8%. This explains some of the difference between US and Australia/New Zealand numbers, but not all.

---

[3] The American conglomerate Tyco International acquired the company in 1990.

TABLE 2.8
Per cent difference between operation of all systems and only wet sprinkler systems

Number of sprinklers operating	United States	Australia and New Zealand		United States	Australia and New Zealand		Difference between total and wet per cent
	Total system per cent	Total system per cent	Difference per cent	Wet system per cent	Wet system per cent	Difference per cent	
1	37.4	64.55	27.2	42.6	64.55	22.0	5.2
2 or fewer	54.6	80.41	25.8	61.0	80.41	19.4	6.4
3 or fewer	63.8	86.54	22.7	70.2	86.54	16.3	6.4
4 or fewer	70.1	89.79	19.7	76.2	89.79	13.6	6.1
5 or fewer	73.4	91.84	18.4	80.2	91.84	11.6	6.8
6 or fewer	77.7	93.44	15.7	83.2	93.44	10.2	5.5
7 or fewer	80.1	94.40	14.3	85.2	94.40	9.2	5.1
8 or fewer	82.2	95.24	13.0	87.0	95.24	8.2	4.8
9 or fewer	83.7	95.79	12.1	88.3	95.79	7.5	4.6
10 or fewer	85.0	96.31	11.3	89.4	96.31	6.9	4.4
11 or fewer	86.2	96.55	10.4	90.4	96.55	6.1	4.2
12 or fewer	87.3	96.82	9.5	91.2	96.82	5.6	3.9
13 or fewer	88.0	97.16*	9.2	91.7	97.16*	5.5	3.7
14 or fewer	89.0	97.51	8.5	92.6	97.51	4.9	3.6
15 or fewer	89.7	97.75	8.1	93.1	97.75	4.7	3.4
20 or fewer	92.2	98.39	6.2	95.0	98.39	3.4	2.8
Average			14.5			9.7	4.8

* This is specified as 96.16% in the original table. This cannot be true and has been changed.

Since the numbers from NFPA are based on the 1970 report covering the period between 1925 and 1969, this book's authors also include the numbers from the 2010 NFPA report (National Fire Protection Association Research, 2010). The authors take the mean value for only wet sprinkler systems from both NFPA reports and compare it to Marryat. This number is probably more relevant to Marryatt's numbers for Australia and New Zealand, as the 2010 NFPA numbers go 17 years longer into the period.

The average difference between countries for only wet systems is 8.1%. The average difference between total mean value for the updated per cent and the per cent for only wet sprinkler systems is 3.6%. This means the updated NFPA numbers only reduce the average gap by 4.8–3.6 = 1.2%.

It does explain some of the difference, but not all.

2.2.2. Unreliability

Marryatt's Chapter 18 on "failure, non-operating sprinkler systems or on ineffectiveness" is called "Fires Not Controlled by Automatic Sprinklers"

TABLE 2.9
Per cent difference between sprinklers operated using updated NFPA numbers and only wet sprinkler systems

Number of sprinklers operating[*]	NFPA reports			Australia			Difference Table 14 vs. updated wet system
	1925–1969	2004–2008		1886–1986			
	Wet System	Wet System	Mean value	Wet system	Difference Table 14	Difference	
1	42.6	52	47.3	64.6	22.0	17.3	4.7
2 or fewer	61	71	66.0	80.4	19.4	14.4	5.0
3 or fewer	70.2	76	73.1	86.5	16.3	13.4	2.9
4 or fewer	76.2	79	77.6	89.8	13.6	12.2	1.4
5 or fewer	80.2	91	85.6	91.8	11.6	6.2	5.4
6 or fewer	83.2	93	88.1	93.4	10.2	5.3	4.9
7 or fewer	85.2	94	89.6	94.4	9.2	4.8	4.4
8 or fewer	87	94	90.5	95.2	8.2	4.7	3.5
9 or fewer	88.3	95	91.7	95.8	7.5	4.1	3.4
10 or fewer	89.4	96	92.7	96.3	6.9	3.6	3.3
20 or fewer	95	97	96.0	98.4	3.4	2.4	1.0
Average					11.7	8.1	3.6

[*] The 2010 report does not have 11–15 or fewer sprinklers.

It was called "Fires in Which Automatic Sprinkler Performance Was Unsatisfactory" in the earlier edition.

The chapter starts with the 1886–1968 period and list 14 cases when the sprinkler system did not operate as expected. Then it lists 23 cases for 1968–1986. It adds 15 fires classified as "Satisfactory" in the earlier edition. The problem is that adding them gives 52 fires that were "not controlled." According to Chapter 4, "Overall Performance Analysis," this should have been 49.

The chapter lists eight causes for 49 fires. Table 2.10. below represents the present authors' attempt to catalogue the 52 separate cases using Marryatt's terminology, but because of missing information in the classification and because of missing cases, the table must be used with caution.

Compared to the reasons for fires given in the NFPA reports, Marryatt's list stands out for several reasons. The first is the naming system. The second is that some of NFPA's definitions incorporate two of Marryatt's reasons into one. For example, "Severe external exposure" and "Unprotected area within or attached to the building" can both be incorporated into the NFPA reason "Fire not in area protected." These fires are not included in the NFPA failure or ineffectiveness analysis, as the fires were outside protected areas.

"System component damaged" consists entirely of fires where automatic extinguishing equipment was damaged by "Explosions" or by ceiling, roof, or building collapse.

TABLE 2.10
Causes of not controlling fires as per cent of separate cases of failure, or ineffective-ness for all structures and wet-pipe sprinklers

Reason	Failure	
	Number	Per Cent
Severe external exposure	5	10%
Unprotected area within or attached to the building	12	25%
Explosions	4	8%
Severity of internal hazard and high fire loading	16	33%
Inadequate water supplies	2	4%
Incendiarism	2	4%
Flash fires and flammable liquids	4	8%
Other factors*	4	8%
Total	**49**	**100%**

* This includes three cases when the sprinkler system was shut off.

"Inappropriate system" can refer to the wrong type of agent, the wrong type of system for an agent, or the wrong design for the system and agent, such as "Severity of internal hazard and high fire loading" and "Flash fires and flammable liquids." "Not enough water discharged" is the same as "Inadequate water supplies."

The overview of Marryatt in Table 2.10. does not include the following four reasons given in the NFPA reports:

1. *Water did not reach fire*: The largest category for ineffectiveness in the NFPA reports is not listed by Marryatt. Typically, this can be the shielding of the sprinkler by obstructions or the shielding of the area where the fire started. For example, case study number 44*, "Rubber Works and Warehouse," is a fire that starts under a temporary cover over a stack of foamed plastic (* authors' numbering). There is no explanation of why this was added. Marryatt does not discuss whether obstruction or shielding could be reasons for the number of sprinklers operated.

2. *Manual intervention*: Two NFPA case studies involve manual inter-vention. Case study number 18*, "Department Store," is a fire where an unauthorized person or persons closed two main stop valves. Case study number 28*, "Furniture Factory," is a fire where the fire brigade shut off the sprinkler system too early (* authors' number-ing). There is no explanation of why they have been added.

3. *Lack of maintenance*: There are no examples of lack of maintenance in the case studies.

4. *System shut off*: This is biggest reason for failure in the NFPA reports. At least seven cases in Marryatt could be classified under this

heading. They are case study numbers 15, 23, 26, 31, 36, 39 and 51* (*
authors' numbering). They are mainly classified as "Incendiarism"
and "Other factors."

At the beginning of Chapter 18, Marryatt says this about systems
shut off: "As in the first edition, the several cases where buildings
and contents were destroyed by fire when the building concerned
were equipped with automatic sprinkler system, but from which
water supplies had been *disconnected permanently, have not been
included in the records*, since these buildings did not in fact have automatic sprinkler system available to operate at the time of the fire"
(*authors' highlight*).

Marryatt does not explain why he has no separate classes for permanently closed systems, such as buildings that are vacant, being
remodelled, or still under construction, or classes for systems that
are temporarily shut off because of system problems like leaks, problems with dirt or water pollution, or damage to pipes or heads.

There seems to be some mixing of terms in Marryatt's chapter. Both
reasons and causes are used. Instead of using the term "System
shut off" as a cause, Marryatt uses the reason the system is shut off.
For example, arsonists shut the system off as part of incendiarism.
Incendiarism is not a cause; it is a reason. In other places, he gives
the cause, for example, "Inadequate water supplies." There can be
many reasons for this, but they are not investigated.

2.2.3. Summary

Marryatt's comprehensive book provides an in-depth look at the 100-year
experience with sprinklers in Australia and New Zealand. However, the
book has flaws. For example, different numbers are given for the same result,
and there are several calculation errors.

Attempts to compare Marryatt's findings for wet sprinklers against newer
NFPA numbers do not explain the big difference between this study and others from around the world.

Nor can we validate Marryatt's findings for reliability. Bukowski (1999)
gives the following explanation of high reliability: "Inspection, testing, and
maintenance exceeded normal expectations, and higher pressures." The systems discussed by Marryatt are wet systems; dry systems have larger fire
growth because of the time water must travel from the sprinkler alarm valve
to the area of fire; furthermore, dry systems have a more inherent possibility
of failure. He also excludes all fires where sprinklers were shut off (except
when arson was involved). Finally, the fact that most of the cases come from
Wormald International Group of Companies, which may have a self-interest
in collecting good reports, and the fact that 99.46% reliability refers to up to
113 sprinklers operating may explain the high reliability he reports.

2.3. Early Studies in the United States

One of the major first studies is a report by the National Fire Protection Association published in the *Fire Journal*, "Automatic Sprinkler Performance Tables, 1970 Edition" (National Fire Protection Association, 1970). This 1970 report has some interesting statistics, findings, tables, and figures; it goes back to 1897 and is one the first reports to present sprinkler performance over a longer period. For example, on page 35, NFPA states: "The tables present below summarize sprinkler performance by occupancy and point out those weaknesses in system *design, installation, and maintenance* that have so far prevented sprinklers from reaching the goal of 100 per cent reliability as a primary means of fire control" (*authors' highlight*). Later, in Table 4 on page 38, NFPA gives more reasons for unsatisfactory performance, for example, the failure of valves, faulty building construction, obstructions, and so on. Whether NFPA thinks 100% reliability can be achieved is not clear.

2.3.1. Reliability

The 1970 NFPA report starts by summarizing sprinkler performance in two columns, the first for 1897 to 1924 and the second for 1925 to 1969. This is reproduced in Table 2.11., below.

TABLE 2.11
Summary of sprinkler performance: 1897–1969

	Fires 1897–1969		Fires 1925–1969*	
	Per cent	Number	Per cent	Number
Satisfactory	95.8%	31.338	96.2%	78.291
Unsatisfactory	4.2%	1.390	3.8%	3.134
Total	100%	32.778	100%	81.425

* "For the five-year period 1965–1969 the ratio of satisfactory performance was 95.7 per cent. As explained in the text, the NFPA now receives fewer reports of favourable performance, relatively, than in previous periods, because small losses are often not reported. Actual sprinkler performance in the five-year period was undoubtedly higher than 95.7 per cent."

Source: Reproduced with permission of NFPA from Automatic Sprinkler Performance Tables, 1970 Edition. Copyright © 1970, National Fire Protection Association, Quincy, MA. All rights reserved.

Of special interest is the term "fire control." On page 25, the 1970 NFPA report states that "sprinkler systems have successfully performed their two main functions—*control and notification*—in 96.2 per cent of the fires" (*authors' highlight*).

Historically, electric fire alarms and sprinklers were used side by side to signal fires. As mentioned earlier, the first practical sprinkler appeared in 1874. The first electric fire sensor to see commercial use was designed by William B. Watkins in 1870 (National Fire Protection Association, u.d.). The

1970 NFPA report does not state how notification of fire in sprinkled buildings is incorporated into the findings.

At the end of page 35, NFPA says: "The word control, as used above, means
prevention of excessive fire spread in light of the nature of the occupancy. In certain occupancies fewer than five sprinklers should establish control, while in
other occupancies over 100 may be needed" (*authors' highlight*).

NFPA 13 (National Fire Protection Association, 2016) categorizes occupancy; for example, Light Hazard, Ordinary Hazard 1, and Ordinary Hazard
2 have the same area of operation (139 m2), and Extra Hazard 1 and Extra
Hazard 2 have a larger area (232 m²). Note that the date when NFPA started
to use hazard classes is unclear.

Since the 1970 NFPA report defines control "in light of the nature of the
occupancy," it would have been natural to present findings based on sprinkler activation within different classifications, but this is not done. Instead,
there is a figure with the cumulative per cent of fires and the numbers of
sprinklers operated.

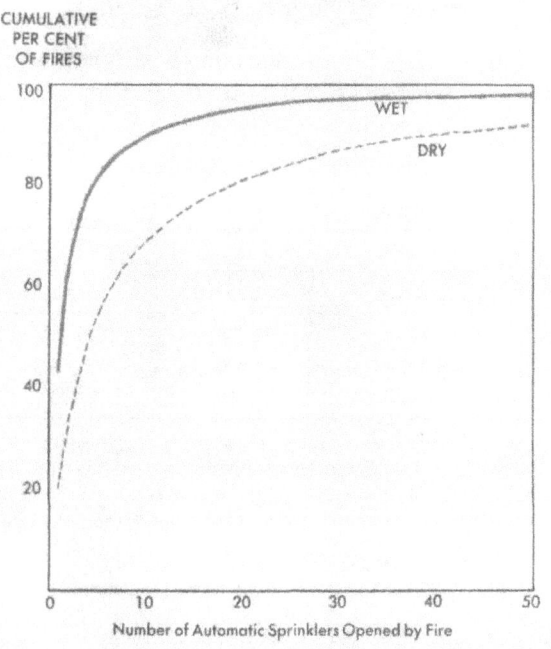

FIGURE 2.1
Wet versus dry-pipe systems, 1925–1969.

In Figure 2.1., the data (taken from the 1970 NFPA report) clearly show
that on average, more dry-pipe sprinklers open than wet-pipe sprinklers; the

delay between the time of tripping the valve and water going through the pipe permits the fire to grow larger. This finding does not support the stated reason for the 1970 NFPA report: to find out if the sprinkler system controls the fire "in light of the nature of the occupancy."

The report goes on to take a more detailed look at performance in different occupancies, like residential, educational, office, and so on, reproduced here in Table 2.12. This is done as a per cent and not as, perhaps expected, the number of sprinklers activated per occupancy.

NFPA's table includes "Total Satisfactory Per Cent" and "Total Unsatisfactory." It is possible to track what "Unsatisfactory" means, as part two of the table shows what constitutes unsatisfactory. It is not possible to track what "Total Satisfactory" means.

"Other Occupancies" in Table 2.12. consist chiefly of idle or vacant buildings, with a 79.2% "Total Satisfactory" rate. The low findings designate buildings where sprinkler maintenance is likely to be substandard. This contrasts with the 88.2% found by Koffel (2006) and the even higher 90.8% found by Budnick (2001). The gap between the 1970 NFPA report and their reports is quite high: 88.2–79.2 = 9% and 90.8–79.2 = 11.6%, for Koffel and Budnick, respectively.

2.3.2. Unreliability

As noted, Table 2.12. is reproduced from the 1970 NFPA report. The table does not differentiate between failed to operate or operating ineffectively under the column "Classification of Unsatisfactory Performance." The table lists 13 main reasons (see Table 4 on page 38 in the NFPA report) for both. It gives four reasons why the system did not work—"Water shut off," "Partial protection," "Inadequate water supply," and "System frozen"—before giving reasons that would probably be classified as not operating effectively, like "Slow Operation", and so on.

So even if there is no formal differentiation, the NFPA author was perhaps thinking about it. The problem is that "Inadequate water supply" is not a reason for not operating; it is a reason for not operating effectively. The following could be classified as "failed to operate."

1. *Water shut off*: If the system is shut off, it can clearly not operate.
2. *Partial protection*: If a system does not cover the whole building, there is no guarantee that there will be a sprinkler where a fire breaks out. A fire that starts outside the protected area will naturally not activate the system (at least at the start), and will perhaps not work effectively.
3. *System frozen*: If a system or part of the system is frozen, there will be no delivery of water.
4. *Defective dry-pipe valve*: Dry valves are more exposed to faults because of their design, but a semi-wet environment is the perfect place for rust. This system has a higher need of maintenance than a wet system. If the valve does not open, the system is, in fact, shut off.

TABLE 2.12
Sprinkler performance summary and classification of unsatisfactory performance

Occupancies	Performance summary	Classification of unsatisfactory performance						
	Total no. Of fires	Total unsatisfactory	Total satisfactory	Total satisfactory per cent	Water shut off	Partial protection	Inadequate water supplies	System frozen
Residential......	1.073	48	1.025	95.5	13	9	5	1
Assembly........	1,551	52	1.499	96.6	23	10	3	...
Educational.....	241	20	221	91.7	4	8	1	...
Institutional.....	305	12	293	96.1	3	3	2	...
Office..........	494	13	481	97.4	4	2	1	...
Mercantile.......	6,237	176	6,061	97.2	83	11	4	4
Industrial								
Beverages, essential oils....	543	64	479	88.2	17	4	9	...
Chemicals......	4.147	198	3.949	95.2	33	11	19	...
Fiber products...	539	25	514	95.3	6	...	4	1
Food products...	2,484	133	2.351	94.6	43	11	8	1
Glass products...	519	23	496	95.6	8	...	3	1
Leather, leather products.....	2.864	114	2.750	96.0	43	8	7	3
Metal, metal products.......	9,807	305	9.502	90.9	91	36	22	3
Mineral products.	394	19	375	95.2	10	4	2	...
Paper, paper products.......	7,147	234	6.913	96.7	75	16	34	3
Rubber, rubber products......	1,489	61	1.428	95.9	21	4	3	...

Slow opera-tion	Defective dry-pipe value	Faulty building construc-tion	Obstruction to distribution	Hazard of occu-pancy	Exposure fire	Inadequate maintenance	Antiquated system	Miscellaneous and unknown
...	...	11	3	1	...	2	2	1
1	...	9	1	...	1	4	...	...
...	...	5	...	...	...	1	1	...
...	...	1	...	1	...	...	...	2
...	1	2	...	1	...	1	1	...
4	5	35	11	12	1	4	1	1
...	1	2	1	18	3	3	5	1
3	3	1	13	95	2	12	1	5
...	2	...	5	4	...	2	1	...
2	1	7	9	29	4	12	1	5
...	...	2	1	5	...	3	...	...
2	4	9	7	9	4	9	6	3
6	6	15	35	43	6	29	7	6
...	...	1	...	...	...	1	1	...
2	2	16	32	21	2	23	4	4
1	1	1	10	14	1	5	...	...

(Continued)

TABLE 2.12 *(Continued)*

	Performance summary	Classification of unsatisfactory performance						
Occupancies	Total no. Of fires	Total unsatisfactory	Total satisfactory	Total satisfactory per cent	Water shut off	Partial protection	Inadequate water supplies	System frozen
Textiles – manufacturing..	16,119	291	15.828	98.2	109	15	32	3
Textiles – processing.	6,527	127	6.400	98.1	52	6	11	...
Wood products..	5,353	492	4.861	90.8	137	57	84	9
Miscellaneous industries......	9,013	265	8.748	97 1	146	15	14	8
Total (industrial)	66,945	2,351	64,594	96.5	791	187	252	32
Storage occupancies	4,160	375	3.785	91.0	122	24	48	5
Other occupancies....	419	87	332	79.2	67	...	...	2
TOTAL (ALL OCCUPANCIES)	81,425	3,134	78,291	96.2	1,110	254	311	44

Since there are no explanations of terms and classifications in the 1970 NFPA report, some reasons for ineffective operation are difficult to interpret.

The report says "Water shut off" is the main cause of ineffective operation in 52% of the cases (1 110 cases out of 2 134).

2.3.3. Summary

The preceding review of the 1970 NFPA report published in the *Fire Journal* points to an overall lack of definitions, explanations, and substantiation of expressions used, for example, "in light of the nature of the occupancy." It would have been natural to present findings based on sprinkler activation within different classifications, but this is not done. Instead, the readers are given a figure that presents a cumulative per cent of fires with the numbers of sprinklers operated.

The focus in this section is obviously the 1970 NFPA report, that Koffel (2006) included into his overview. However, this book's authors note that it is

Slow opera-tion	Defective dry-pipe value	Faulty building construc-tion	Obstruction to distribution	Hazard of occu-pancy	Exposure fire	Inadequate maintenance	Antiquated system	Miscellaneous and unknown
5	3	11	27	18	1	50	9	8
5	1	8	13	15	2	7	1	6
16	14	27	19	77	8	24	12	8
3	. . .	12	11	18	3	27	12	8
45	38	112	183	366	36	207	56	46
6	9	10	57	38	11	40	3	7
. . .	. . .	2	1	5	3	3	1	3
56	53	187	256	424	52, (262). 65. 60			

not clear why Koffel uses the second lowest individual performance rate of 88.2% and not the lowest, 79.2%, without remarking on his choice in any way.

2.4. US Experience With Sprinklers: 2010

In September 2010, NFPA's Fire Analysis and Research Division released "U.S. Experience with Sprinkler and Other Automatic Fire Extinguishing Equipment" (National Fire Protection Association Research, 2010). As the 2010 NFPA report makes clear, there have been many changes since the Automatic Sprinkler Performance Tables were published in the *Fire Journal* in 1970 (National Fire Protection Association, 1970). Some of the changes are obvious; for example, the earlier article has five pages and the more recent one has 87. More importantly, NFPA has made major changes in methodology and presentation.

1. The data used in this report not only come from NFPA's databank, but also from the detailed information available in Version 5.0 of the US Fire Administration's National Fire Incident Reporting System (NFIRS 5.0). These fires are reported by the US municipal fire departments, so the report excludes fires reported only to federal or state agencies or industrial fire brigades.

2. NFIRS 5.0 was introduced in 1999 and brought major changes to fire incident data, including changes in definitions and coding rules. There were further changes in 2003. Data for 1999–2003 are not used in the 2010 report. This report is for 2004–2008.

3. As Appendix A in the report (National Fire Protection Association Research, 2010) states, the NFIRS is voluntary. Roughly two-thirds of the US municipal fire departments participate. To address this issue and to update the NFPA Fire Records, the annual NFPA Fire Department Experience is sent to all municipal departments protecting populations of 50 000 or more. The study is used to project national estimates by comparing NFPA's projected totals to data in NFIRS. Even though there are uncertainties in this method, it gives a far better picture from an analytical point of view.

Projection is based on the following equation:

$$\frac{NFPA\,survey\,projections}{NFIRS\,totals\,(Version\,5.0)}$$

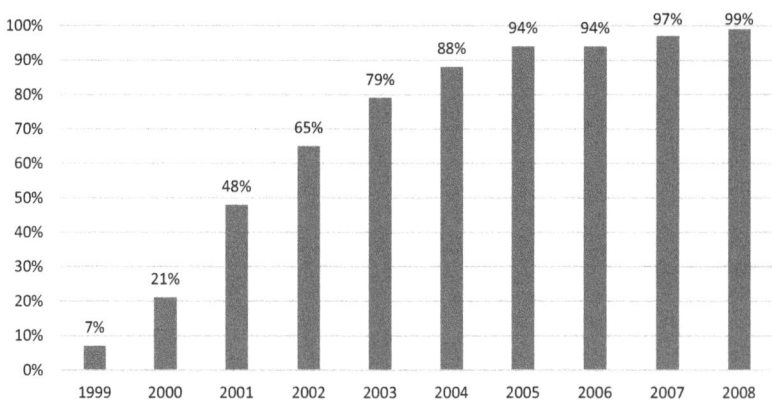

FIGURE 2.2
Fires originally collected in NFIRS 5.0 by year: 1999–2008.

Source: Reproduced with permission of NFPA from "U.S. Experience with Sprinkler and Other Automatic Fire Extinguishing Equipment," 2010. Copyright © 2010, National Fire Protection Association, Quincy, MA. All rights reserved.

4. Appendix B in the 2010 NFPA report (National Fire Protection Association Research, 2010) shows a systematic approach to reporting on automatic extinguishing systems.

A. **M1 Presence of Automatic Extinguishment System**

   N None Present

   1 Present

   U Undetermined

B. **M2 Type of Automatic Extinguishment System**

   1 Wet-pipe sprinkler

   2 Dry-pipe sprinkler

   3 Other sprinkler system

   4 Dry chemical system

   5 Foam system

   6 Halogen-type system

   7 Carbon dioxide system

   0 Other special hazard system

   U Undetermined

C. **M3 Automatic Extinguishment System Operation**

   1 System operated and was effective

   2 System operated and was not effective

   3 Fire too small to activate the system

   4 System did not operate

   0 Other

   U Undetermined

D. **M4 Number of Sprinklers Operating**

E. **M5 Automatic Extinguishment System Failure Reason**

   1 System shut off

   2 Not enough agent discharged [to control the fire]

   3 Agent discharged but did not reach [the] fire

   4 Wrong type of system [inappropriate system for the type of fire]

   5 Fire not in area protected [by the system]

   6 System components damaged

   7 Lack of maintenance [including corrosion or heads painted]

   8 Manual intervention [defeated the system]

   0 Other _____ [other reason system not effective]

   U Undetermined

It seems that information about design requirements of automatic extinguishment systems is not collected.

2.4.1. Reliability

In the Fact Sheet on page vii, the 2010 NFPA report states: "In reported structure fires large enough to activate them, sprinklers operated in 91% of fires in sprinklered properties." Furthermore, "In reported structure fires large enough to activate them, sprinklers operated and were effective in 87% of fires in sprinklered properties." These findings are shown in Table 3 in the 2010 report based on property use and are reproduced here in Table 2.13.

Table 2.13. indicates that 49% of all fires to which the fire department responded were too small to activate the extinguishing system. When an extinguishing system was present, the fire was large enough to activate the equipment in 51% of the cases; sprinklers were present in the fire area in 91% of the cases, and in 87% of these, the equipment operated effectively.

The 2010 NFPA report does not have definitions, nor does it explain how key questions were answered.

Example 1: "Per Cent of Fires Too Small to Activate Equipment." How small is too small? Presumably a smouldering fire would, in many cases, not give off enough heat to activate a thermal bulb on a sprinkler head, but this is not defined, and there is no explanation of how the key findings were derived. This could also be applied to the Omega sprinkler failure; another study reports that roughly one-third of Omega sprinklers failed to operate under the required pressure (Fire Engineering, 1997). Failure should be of interest for the fire community in general, including fire departments and regulatory agencies alike.

Example 2: "Equipment Operated Effectively." What is effective? Is this fire control without manual interventions from residents, employers, or fire departments? A hint is given on page 17:

> As noted, for most rooms in most properties, *effective performance is indicated by confinement of fire to the room of origin*. For the few rooms where the design area is smaller than the room, a sprinkler system can be ineffective in terms of confining fire to the design area but still be successful in confining fire to the larger room of origin. Therefore, one might expect the percentage of fires with flame confined to room of origin to be slightly larger than the combined performance (operating effectively) for any given property use. Table B shows this is usually the case (*authors' highlight*).

According to page 15 in the 2010 NFPA report (National Fire Protection Association Research, 2010), "*Effectiveness should be measured relative to the design objectives for a particular system. For most rooms in most properties, sprinklers are designed to confine fire to the room of origin.*" (*authors' highlight*). It seems there are some assumptions on room and fire spread. See 3.3.2. of this book, "How to Collect Data," for more information on this.

TABLE 2.13

Automatic extinguishing equipment reliability and effectiveness, by property use for 2004–2008 structure fires (excluding fires reported as confined fires)

A. All sprinklers

Property use	Number of fires per year where extinguishing equipment was present	Per cent of fires too small to activate equipment	When equipment was present, fire was large enough to activate equipment, and sprinklers were present in fire area			
			Number of fires per year	Per cent where equipment operated (A)	Per cent effective of those that operated (B)	Per cent where equipment operated effectively (A x B)
All public assembly	1 350	50%	680	89%	92%	82%
Eating or drinking establishment	770	46%	410	90%	90%	81%
Educational property	810	71%	240	85%	96%	82%
Health care property*	1 320	69%	400	87%	97%	84%
Residential	6 760	44%	3 790	94%	97%	91%
Home (including apartment)	4 860	38%	3 000	94%	97%	92%
Hotel or motel	810	60%	330	91%	98%	89%
Dormitory or barracks	260	62%	100	91%	99%	90%
Rooming or boarding house	210	47%	110	93%	96%	90%
Board and care home	170	60%	70	91%	97%	89%
Store or office	2 590	54%	1 200	89%	97%	86%
Grocery or convenience store	510	60%	210	88%	95%	83%
Laundry or dry cleaning	240	47%	130	91%	95%	87%
Service station or motor vehicle sales or service	110	32%	80	93%	94%	87%

(Continued)

TABLE 2.13 (*Continued*)

A. All sprinklers

Property use	Number of fires per year where extinguishing equipment was present	Per cent of fires too small to activate equipment	When equipment was present, fire was large enough to activate equipment, and sprinklers were present in fire area			
			Number of fires per year	Per cent where equipment operated (A)	Per cent effective of those that operated (B)	Per cent where equipment operated effectively (A x B)
Department store	370	61%	150	88%	98%	86%
Office	520	62%	200	89%	97%	86%
Manufacturing facility	2 470	42%	1 420	90%	93%	84%
All storage	600	35%	390	80%	96%	76%
Warehouse excluding cold storage	340	34%	230	85%	97%	82%
All structures**	**16 600**	**49%**	**8 430**	**91%**	**96%**	**87%**

* Nursing home, hospital, clinic, doctor's office, or development disability facility.

** Includes some properties not listed separately above.

Source: Reproduced with permission of NFPA from "U.S. Experience with Sprinkler and Other Automatic Fire Extinguishing Equipment," 2010. Copyright © 2010, National Fire Protection Association, Quincy, MA. All rights reserved.

Example 3: According to pages 16–17 and Table 6 in the 2010 NFPA report, reproduced as Table 2.14. in this book, "Table 6 [Table 2.14.] provides direct measurement of sprinkler effect involving the first bulleted scenario on the previous page. For all structures combined, 73% have flame damage confined

TABLE 2.14

Extent of flame damage, for sprinklers present vs. automatic extinguishing equipment absent for 2004–2008 structure fires

Property use	Per cent of fires confined to room of origin excluding structures under construction and sprinklers not in fire area		
	With no automatic extinguishing equipment	With sprinklers of any type	Difference (in percentage) points
Public assembly	76%	95%	19
Fixed-use amusement or recreation place	75%	96%	21
Variable-use amusement or recreation place	84%	97%	13
Religious property	72%	96%	24
Library or museum	83%	97%	14
Eating or drinking establishment	75%	94%	19
Educational	90%	98%	8
Health care property*	93%	99%	6
Residential	76%	97%	21
Home (including apartment)	76%	97%	21
Hotel or motel	86%	97%	11
Dormitory or barracks	96%	99%	3
Store or office	71%	93%	22
Grocery or convenience store	76%	96%	20
Laundry or dry cleaning or other professional supply or service	80%	92%	12
Service station or motor vehicle sales or service	61%	88%	27
Department store	73%	93%	20
Office building	76%	94%	18
Manufacturing facility	69%	86%	17
Storage	32%	80%	48
Warehouse excluding cold storage	53%	81%	28
All structures**	**73%**	**95%**	**22**

* Nursing home, hospital, clinic, doctor's office, or development disability facility.
** Includes some properties not listed separately above.

Source: Reproduced with permission of NFPA from "U.S. Experience with Sprinkler and Other Automatic Fire Extinguishing Equipment," 2010. Copyright © 2010, National Fire Protection Association, Quincy, MA. All rights reserved.

to room of origin when there is no automatic extinguishing equipment present. This rises to 95% of fires with flame damage confined to room of origin when any type of sprinkler is present."

For 27% of the fires with no sprinkler system present, the fire goes beyond the room of origin. The number is 5% for fires with a sprinkler system. Is a 95/73 = 30% increase in reducing fire spread when a sprinkler system is present scientific proof of success for a sprinkler system, or is it something else? Perhaps it proves that fire barriers (every room is a fire barrier) together with sprinkler systems can prevent the spread of fire in 95% of the cases, even if the design area is smaller than the room in question.

Example 4: According to page 18 of the 2010 NFPA report, "*Effectiveness declines when more sprinklers operate.* When more than 1–2 sprinklers have to operate, this may be taken as an indication of less than ideal performance" (*authors' highlight*). Later, the report says: "At the same time, the number of sprinklers operating should not be used as an independent indicator of effectiveness because sprinklers are deemed effective in most fires where sprinklers operate, no matter how many sprinklers operate." It is not clear what the NFPA author means.

Example 5: Table 13 in the NFPA report indicates that 49% of all fires to which fire departments responded were too small to activate the extinguishing system: "Per Cent of Fires Too Small to Activate Equipment." However, it should be possible to only refer to sprinklers and not to extinguishing systems more generally, when it is the findings about sprinklers that are of interest.

Furthermore, there is no definition of a fire "large enough" or a fire "too small to activate" the sprinkler. This is one of the most important findings in this report, but it is not treated or discussed as one would expect. Of course, small fires can activate smoke detectors, fire alarms may be manually activated alerting the fire department, people may intervene, or fires may be smouldering and not trigger the sprinkler system. The unanswered question is: how does the person reporting the fire know when the system should activate?

2.4.2. Unreliability

As noted at the beginning of this chapter, by 2010, NFPA had made a major shift in methodology and presentation. One is a clear differentiation between "failed to operate" or "operated ineffectively," as stated above. What are the reasons for failure to activate? Table 2.15. is taken from the 2010 NFPA report.

As the table shows, in 2010, NFPA had five possible reasons for not operating. NFPA's 1970 report (National Fire Protection Association, 1970) gives four reasons for not operating: "Water shut off," "Partial protection," "System frozen," and "Defective dry-pipe valve." The 2010 report gives five reasons, only one of which is from the original report: "Water shut off." The five reasons listed in the 2010 report are the following:

TABLE 2.15

Reasons for failure to operate when fire was large enough to activate equipment and equipment was present in area of fire, by property use based on indicated estimated number of 2004–2008 structure fires per year (excluding fires reported as confined fires)

A. All sprinklers

Property use	System shut off	Manual intervention defeated system	Lack of maintenance	Inappropriate system for type of fire	System component damaged	Total fires per year
All public assembly	61%	14%	12%	10%	2%	74
Eating or drinking establishment	64%	15%	21%	0%	0%	41
Residential	54%	20%	9%	9%	7%	234
Home (including apartment)	57%	15%	9%	11%	9%	167
Store or office	62%	20%	8%	6%	3%	131
Manufacturing facility	64%	17%	7%	4%	7%	141
Storage	84%	5%	5%	1%	4%	79
All structures*	**64%**	**17%**	**8%**	**6%**	**5%**	**100%**
All structures	**513**	**136**	**64**	**48**	**40**	**801***

* Includes some properties not listed separately above.

Source: Reproduced with permission of NFPA from "U.S. Experience with Sprinkler and Other Automatic Fire Extinguishing Equipment," 2010. Copyright © 2010, National Fire Protection Association, Quincy, MA. All rights reserved.

1. *System shut off*: If the system is shut off, it can clearly not operate. This is called Water Shut Off in the report (National Fire Protection Association, 1970), but it includes, for example, pumps that are shut off/out of order.

2. *Manual intervention*: With a fire in the building, is very important that the system is shut off at the right time. Intervention should always be the responsibility of the fire department.

3. *Lack of maintenance*: It is important that valves, pipes, and sprinklers are maintained correctly according to standards and recommended procedures, so that the intended function is maintained.

4. *Inappropriate system*: "Inappropriate System" can refer to the wrong type of agent (e.g. water vs. chemical agent or carbon dioxide), the wrong type of system for the same agent (e.g. wet pipe vs. dry pipe), or the wrong design for the system and agent (e.g. a design adequate only for Class I commodities vs. a design adequate for any class of commodities). It is not clear how this is determined or how a fire in a room with a chemical agent or incorrect design will affect whether the system works or not.

5. *System component damaged*: "In the NFPA compilation of incidents of failure or ineffectiveness, the incidents involving component damage consist entirely of fires where automatic extinguishing equipment was damaged by explosions or by ceiling, roof, or building collapse, *nearly always as a consequence of fire*" (authors' highlight). Except for explosions (that could start a fire), this occurs after a fire breaks out. How can a fire that has already started and results in the collapse of a roof or a ceiling be the reason for the damaged component causing the system not to work? Did the system not respond to the fire, or did the fire start outside the sprinkled area? This is not clear.

   The following two tables, Table 2.16. and Table 2.17., are based on Table 4 and Table 5, respectively, in the 2010 NFPA report. **Table 2.16** shows the combined failure and ineffective sprinkler performance ("Water discharged but did not reach fire" and "Not enough water discharged").

As Table 2.16. shows, there is a drop in per cent of "System Shut Off", from 52% in the 1970 report to 45% in the 2010 report, where it is combined with ineffectiveness.

Alternatively, the per cent of failure and ineffectiveness can be presented as separate cases of failure and ineffectiveness per year, as shown in Table 2.17.

2.4.3. Summary

Compared to the 1970 NFPA article, the 2010 NFPA report is very comprehensive. There has obviously been a major shift in methodology and presentation. NFIRS is now the main source of data. The report's tables give

TABLE 2.16
Reasons for failure or ineffectiveness as number of fires per year and percentages of all cases of failure or ineffectiveness, for all structures and all types of sprinklers

Reason	Failure		Ineffectiveness		Combined	
	Number[*]	Per Cent	Number[*]	Per Cent	Number[*]	Per Cent
System shut off	513	45%	0	0%	513	45%
Manual interruption defeated system	136	12%	23	2%	159	14%
Water discharged but did not reach fire	0	0%	144	13%	144	13%
Not enough water discharged	0	0%	88	8%	88	8%
Lack of maintenance	64	6%	26	2%	90	8%
Wrong type of (inappropriate) system for type of fire	48	4%	20	2%	68	6%
System component damaged	40	4%	26	2%	66	6%
Total	**801**	**71%**	**327**	**29%**	**1 128**	**100%**

[*] The numbers are not given in the table but calculated as per cent of the given total number.

Source: Reproduced with permission of NFPA from "U.S. Experience with Sprinkler and Other Automatic Fire Extinguishing Equipment," 2010. Copyright © 2010, National Fire Protection Association, Quincy, MA. All rights reserved.

TABLE 2.17
Reasons for failure or ineffectiveness as per cent of separate cases of failure or ineffectiveness, for all structures and all type sprinklers

Reason	Failure		Ineffectiveness		Combined	
	Number[*]	Per Cent	Number[*]	Per Cent	Number[*]	Per Cent
System shut off	513	64%	0	0%	513	45%
Manual interruption defeated system	136	17%	23	7%	159	14%
Water discharged but did not reach fire	0	0%	144	44%	144	13%
Not enough water discharged	0	0%	88	27%	88	8%
Lack of maintenance	64	8%	26	8%	90	8%
Wrong type of (inappropriate) system for type of fire	48	6%	20	6%	68	6%
System component damaged	40	5%	26	8%	66	6%
Total	**801**	**100%**	**327**	**100%**	**1 128**	**100%**

[*] The numbers are not given in the table but calculated as per cent of the given total number.

Source: Reproduced with permission of NFPA from "U.S. Experience with Sprinkler and Other Automatic Fire Extinguishing Equipment," 2010. Copyright © 2010, National Fire Protection Association, Quincy, MA. All rights reserved.

information on reliability and effectiveness by property use, reasons for failure to operate, and ineffectiveness.

However, there are some problems, including a lack of definitions and not enough explanation of the assumptions. For example, the following are not explained: "fires large enough to activate them" and "the nature of the occupancy."

2.5. US Experience With Sprinklers: 2017

In July 2017, NFPA Fire Research released "U.S. Experience with Sprinklers" (National Fire Protection Association Research, 2017b) for 2010–2014.

The methodology from the 2010 report is continued in the 2017 report. However, 100% of the data now come from the NFIRS databank. Now that all the data come from the most recent NFIRS, the numbers should be of greater interest, but they have been scaled by an unknown ratio based on the NFPA annual Fire Department Experience Survey. How much influence does it have on the numbers?

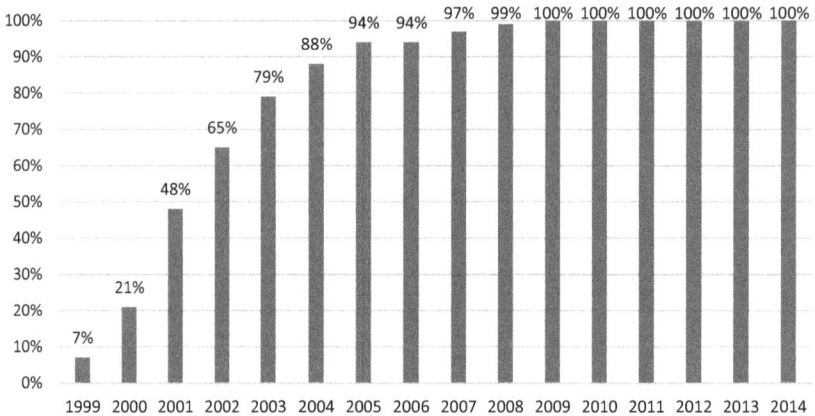

FIGURE 2.3
Fires originally collected in NFIRS 5.0 by year: 1999–2014.

Source: Reproduced with permission of NFPA from "U.S. Experience with Sprinklers, 2017." Copyright © 2017, National Fire Protection Association, Quincy, MA. All rights reserved.

In addition to estimating number of fires, fatalities, and fire losses, the 2017 report looks at sprinkler operation and effectiveness. The report says:

> All estimates in this report exclude fires in properties under construction. Fires in which partial systems were present and fires in which sprinklers were present but failed to operate because they were not in the fire area were excluded from estimates related to presence and operation.

(p. 1)

"Confined fires" are excluded from this report and the previous report. A confined fire is a fire within a chimney or flue, fuel burner or boiler, cooking vessel, incinerator, commercial compactor, or trash. The report also excludes automatic fire extinguishing systems other than sprinklers and goes into depth on the numbers of home fires.

2.5.1. Reliability

On page 5, the 2017 NFPA report states: "Sprinklers operated in 92% of the fires in which sprinklers were present and the fire was considered large enough to activate them. They were effective at controlling the fire in 96% of fires in which they operated. Figure 8 [see Table 2.18. in this text] shows that sprinklers operated effectively in 88% of the fires large enough to trigger them."

TABLE 2.18
Automatic sprinkler system reliability and effectiveness, by property use for 2010–2014 structure fires (excluding fires reported as confined fires)

A. All sprinklers

Property use	Number of fires per year where sprinkler was present	Property use	When equipment was present, fire was large enough to activate equipment, and sprinklers were present in fire area		
			Number of fires per year	Per cent where sprinkler operated (A)	Per cent effective of those that operated (B)	Per cent where sprinkler operated effectively (A x B)
All public assembly	1 220	48% (580)	640	90%	94%	85%
Eating or drinking establishment	710	46% (300)	410	90%	92%	83%
Educational property	590	70% (410)	180	87%	96%	84%
Health care property*	900	66% (590)	310	87%	97%	84%
Residential	6 630	38% (2 490)	4 140	93%	96%	89%
Home (including apartment)	5 470	35% (1 900)	3 570	94%	96%	91%
Hotel or motel	680	52% (350)	330	90%	98%	89%
Store or office	2 070	50% (1 030)	1 040	91%	96%	87%
Grocery or convenience store	430	56% (240)	190	89%	93%	83%

(Continued)

TABLE 2.18 *(Continued)*

A. All sprinklers

Property use	Number of fires per year where sprinkler was present	Property use	When equipment was present, fire was large enough to activate equipment, and sprinklers were present in fire area			
			Number of fires per year	Per cent where sprinkler operated (A)	Per cent effective of those that operated (B)	Per cent where sprinkler operated effectively (A x B)
Department store	270	56% (150)	120	90%	98%	88%
Office	400	55% (220)	180	91%	96%	87%
Manufacturing facility	1 360	44% (600)	1 030	91%	94%	85%
All storage	440	32% (140)	300	86%	96%	82%
Warehouse excluding cold storage	270	33% (90)	180	84%	97%	81%
All structures*	**13 210**	**44% (5 840)**	**7 640**	**92%****	**96%****	**88%****

* Nursing home, hospital, clinic, doctor's office, or development disability facility.
** The per cent is taken from Table 6 in the 2017 NFPA report. Note: the per cent has not been checked.

Table 2.18. is taken directly from the 2017 report. It shows the automatic sprinkler system reliability and effectiveness by property use for 2010–2014 structure fires (excluding fires reported as confined fires). Table 2.13. (see earlier), taken from the 2010 report (National Fire Protection Association Research, 2010), sets the per cent of fires too small to activate equipment at 49%, but by 2017, as Table 2.18. shows, the per cent has dropped to 44%. The numbers in the 2010 report include all extinguishing equipment, but this newer report only looks at sprinklers; this could be the reason for the drop in the numbers of too small fires from 16 600 to 13 210.

Unfortunately, seven of 14 numbers in the column "Non-confined fires too small to activate or unclassified operation" in the original Table 6 do not match. The first example, "All public assembly," reports there were 3 760 fires where a sprinkler was present and classifies 2 540 of these as "Confined fires." This means that for 3 670–2 540 = 1 220 non-confined fires, a sprinkler was present. If 640 fires were coded as large enough to activate sprinklers, then 1 220–640 = 580 were too small to activate sprinklers. The problem is that Table 6 in the 2010 report states that 590 fires were too small to activate sprinklers or were unclassified operations.

As shown in Figure 2.4., out of five fires, only one head activated in four cases (79%).

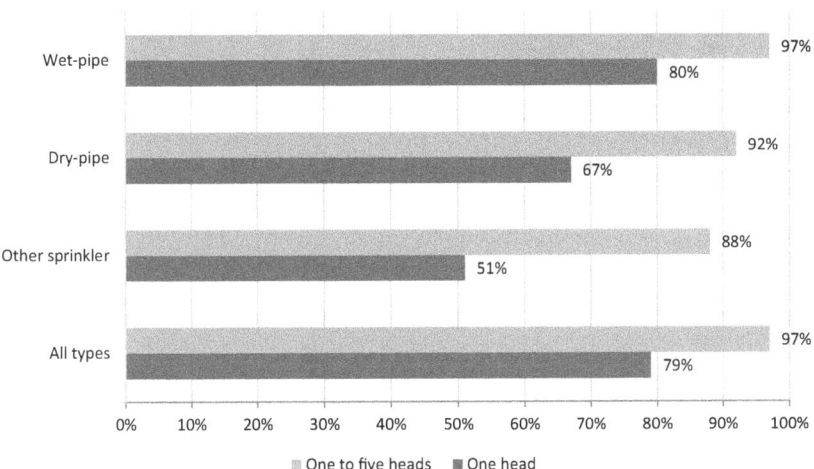

Some types of sprinklers were present in an average of 49 840 fires per year (including confined fires). These fires caused an annual average of 42 deaths, representing 2% of all fire deaths. Figure 2.5. shows in per cent the difference in death rates for buildings where a sprinkler was present and those without an AES.

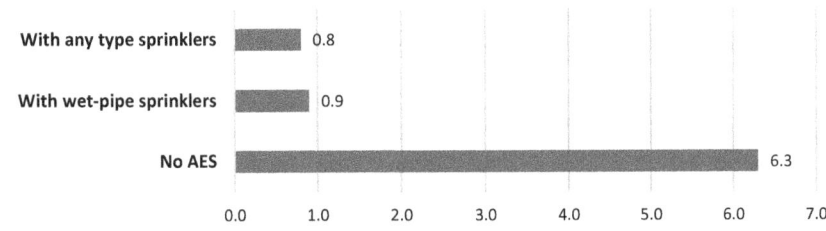

Sprinklers were present in an average of 24 440 home fires per year. These fires caused an average of 35 deaths per year. One per thousand fires resulted in deaths in houses with sprinklers; seven per thousand deaths occurred in houses without sprinklers. According to the 2017 NFPA report, the number of deaths was 81% lower when sprinkler systems were present than when there were no automatic extinguishing systems (see Figure 2.6.).

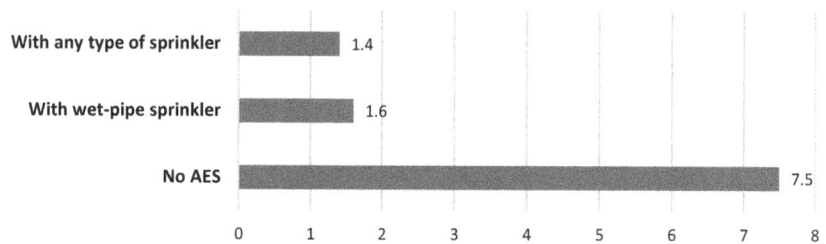

FIGURE 2.6
Civilian death rates per 1 000 fires in homes with sprinklers and with no AES: 2010–2014.

Source: Reproduced with permission of NFPA from "U.S. Experience with Sprinklers, 2017." Copyright © 2017, National Fire Protection Association, Quincy, MA. All rights reserved.

After looking at deaths and injuries, the 2017 report shifts to sprinkler operation and effectiveness in home fires. As stated earlier and as shown in Figure 2.7., the general operating effectivity was 88%, and for home fires, it was 91%.

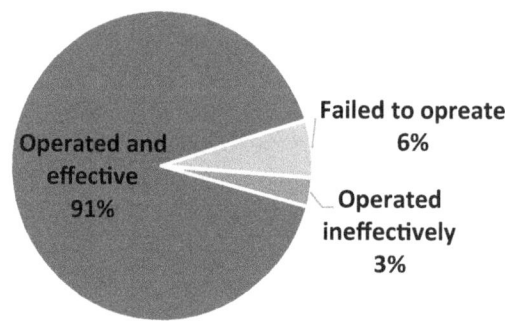

FIGURE 2.7
Sprinkler operation and effectiveness in home fires: 2010–2014.

Source: Reproduced with permission of NFPA from "U.S. Experience with Sprinklers, 2017." Copyright © 2017, National Fire Protection Association, Quincy, MA. All rights reserved.

There are no definitions in the 2017 report, and it is not clear if homes are separated from other residential structures. Page 2 of the report's "Fact Sheet" states: "Although the majority of structure fires, civilian fire deaths and injuries, and property damage occurred in residential properties, *particularly homes*, only 8% of the reported residential fires were in properties with sprinklers" (*authors' highlight*). Page 3 of the "Fact Sheet" has the following explanation: "Homes include one- or two-family homes and apartments or other multi-family homes." Whether this means, for example, that nursing homes are not homes, but *are* residential, is not clear.

The NFPA chapter "Sprinklers in Home Fires" says sprinklers are found in 7% of homes, compared to 12% in all types of buildings. To this, it adds, "*In 98% of home fires with operating sprinklers, five or fewer heads operated*" (p. 12). The (normal) design numbers of residential sprinklers is up to two and four for the biggest rooms, and four for the most hydraulically demanding areas, depending on type of standard (see 3.3.2. in this book, "How to Collect Data"). It is not clear if the reported sprinklers were only residential sprinklers or included other types. Nor is it clear what the purpose is for including five sprinkler heads, when normal design criteria typically include fewer for residential buildings/homes.

What is clear is that in 88% of the cases, only one head activated; see Figure 2.8.

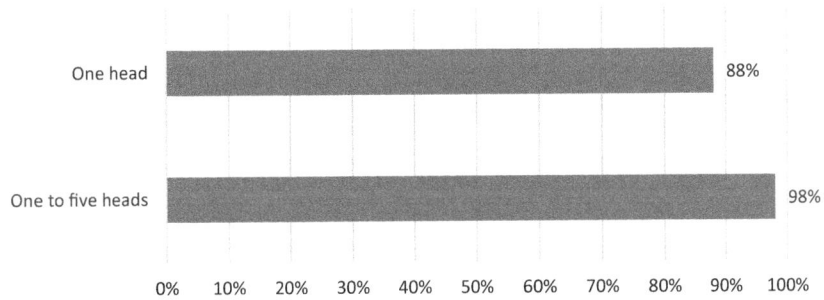

FIGURE 2.8
When sprinklers operated, percentage of home fires in which one or one to five heads operated: 2010–2014.

Source: Reproduced with permission of NFPA from "U.S. Experience with Sprinklers, 2017." Copyright © 2017, National Fire Protection Association, Quincy, MA. All rights reserved.

There is also a lack of clarity in the following:

1. *Too small vs. large enough*: The report does not supply objective criteria for size.
2. *Sprinkler operated effectively*: As noted above, the design criteria are unclear.
3. *Confined to room with start fire*: Fires confined to room of origin represent 96% with sprinklers and 71% with no AES. It is not clear if this is a success criterion.

2.5.2. Unreliability

According to Figure 2.9. (Figure 13 in the 2017 NFPA report), there is a drop in "System shut off" from 52% in the 1970 report (National Fire Protection Association, 1970), to 45% in the 2010 report (National Fire Protection Association Research, 2010), to 40% in this report where it is combined with ineffectiveness.

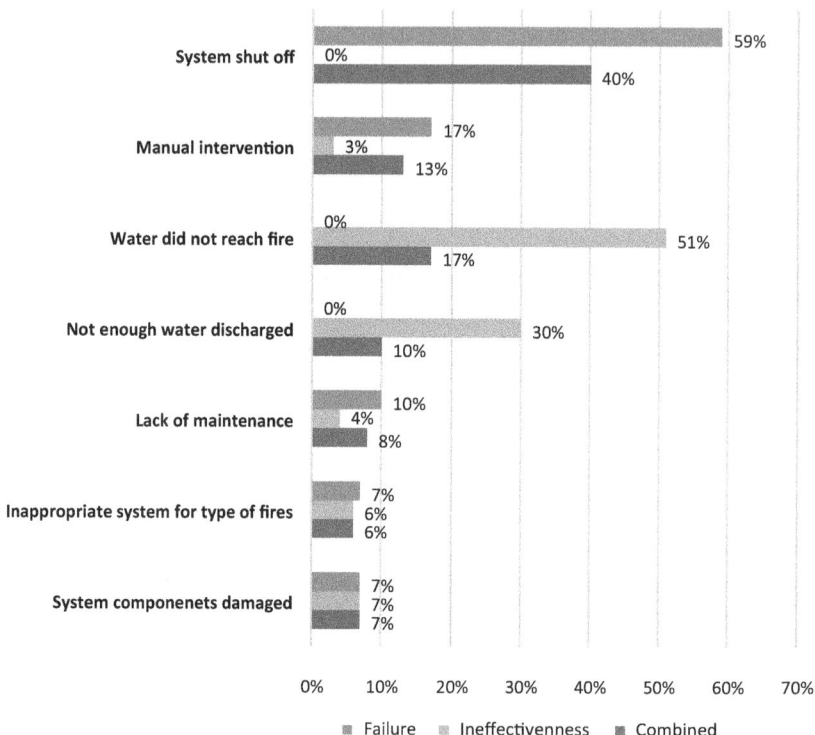

FIGURE 2.9
Reasons for combined sprinkler failure and ineffectiveness: 2010–2014.

Source: Reproduced with permission of NFPA from "U.S. Experience with Sprinklers, 2017."
Copyright © 2017, National Fire Protection Association, Quincy, MA. All rights reserved.

When looking at the bar graph, it is hard to see the differences between the 2010 reports table and the 2017 reports graph (National Fire Protection Association Research, 2010). Therefore, the authors of the present book compare data in the 2010 report's "Reasons for failure or ineffectiveness as percentages of separated cases of failure or ineffectiveness, for all structures and all type sprinklers" (included in this book as Table 2.17) with the data from the 2010–2014 bar graph in the 2017 report (included in this book as Figure 2.9.) to create Table 2.19.

When the columns "Ineffectiveness" and "Combined" are summed for 2010–2014, the total passes 100%, suggesting the rounding is incorrect somewhere. What is important is that the "System shut off" percentage continued to drop and the "Water did not reach fire" percentage increased.

Long-term trends and changes over a shorter time are not included in the 2017 report, representing an important missing tool for fire engineers, fire departments, and others concerned with fire safety. There also seems to be some mixing of terms in the 2017 NFPA report. Both reasons and causes are used.

TABLE 2.19

Comparing reasons for failure or ineffectiveness as percentages of separate cases of failure or ineffectiveness from 2004–2008 to 2010–2014

Reason	Failure		Ineffectiveness		Combined	
Year	2004–2008	2010–2014	2004–2008	2010–2014	2004–2008	2010–2014
System shut off	64%	59%	0%	0%	45%	40%
Manual intervention*	17%	17%	7%	3%	14%	13%
Water did not reach fire*	0%	0%	44%	51%	13%	17%
Not enough water discharged	0%	0%	27%	30%	8%	10%
Lack of maintenance	8%	10%	8%	4%	8%	8%
Inappropriate system for type of fire*	6%	7%	6%	6%	6%	6%
System component damaged	5%	7%	8%	7%	6%	7%
Total	100%	100%	100%	101%	100%	101%

* The names are not the same as in the 2010 report.

2.5.3. Summary

In the 2017 NFPA report, 100% of the data now come from the NFIRS databank, but have been scaled by an unknown ratio based on the NFPA annual Fire Department Experience Survey.

The major focus is on fires in homes, and the probability of death is over eight times higher in a home without AES, than in a home with sprinklers. The report does not explain why it focuses on five or fewer operated sprinklers, when the design for residential buildings is at least two and up to four.

No long-term trends and changes are included in the 2017 report, but the authors of the present book have incorporated them in the review.

2.6. UK Experience With Sprinklers: 2017

Table 1.1. in this book mentions a study by M. Finucane and D. Pinckney (1988), titled "Reliability of Fire Protection and Detection Systems." As noted in the table, the authors of this book have only been able to find Finucane and Pinckney (1987), "Reliability of Fire Protection and Detection Systems: Recent Developments in Fire Detection and Suppression Systems." In the edition that the authors found, there is no mention of 96.9–97.9% reliability for sprinkler systems in the United Kingdom. There is a reference to a study by the Home Office where sprinkler reliability is given as 95%. The authors do not know if there are two reports from Finucane and Pinkney with the same name, or if there is only one, with a first edition and a second more extended edition, or if one of the two dates are an error.

The report "Efficiency and Effectiveness of Sprinkler Systems in the United Kingdom: An Analysis from Fire Service Data" (Optimal Economics, 2017) was prepared for the Chief Fire Officers Association (CFOA) and the National Fire Sprinkler Network (NFSN) by Optimal Economics[4] (OE). OE analyzed data on the activation and performance of sprinkler systems to control fires in buildings. Data were collected from 47 of 52 fire and rescue services through the Incident Recording System (IRS) in the UK by CFOA and NFSN, for the years 2011 to 2015/16. The three fire and rescue services reported no fires with sprinkler systems. Some fire and rescue services reports are by financial year (2011/12 to 2015/16), but most are by calendar year. They have been adjusted to 2011 to 2015. The OE report looks at data and analysis framework, and analysis and results.

2.6.1. Reliability

From 2011 to 2015 (excluding 26 fires from January to March 2016), there were 2,268 fires with sprinkler systems. Most of the fires, 75%, were in non-residential buildings, and 18% were in dwellings.

The summary of the OE report, point 3, says: "The aim of the analysis was to provide an authoritative assessment of the *reliability and effectiveness* of sprinkler systems in controlling and extinguishing fires and in preventing damage" (*authors' highlight*). There are no definitions in the report.

Point 4 says: "The *effectiveness and reliability* of sprinklers has been assessed with regard to two key criteria:

- When sprinklers operate how effective are they in extinguishing or controlling fires and thus preventing damage? (*performance effectiveness*) . . ., the performance effectiveness of sprinkler systems was 99% across all building types.

- How reliable are sprinklers in coming into operation when a fire breaks out? (*operational reliability*)" (*authors' highlight*).

The OE report's conclusion is: "This indicates that the operational reliability of the system was 94%."

There seems to be an assumption that efficiency (title of the report), reliability (point 3 in the summary), and operational reliability (point 4) are the same. The report does not explain why they are the same or why there is a need to use three different words for the same thing.

---

[4] Optimal Economics is a UK based analyst firm in economics, financial appraisal, and policy.

Of the 945 cases where sprinklers operated, data are only available for 532 cases. There is no explanation of whether the 532 cases are representative of all the cases. Figure 2.10. (Figure 5 in the OE report) shows how often sprinklers were in the room of fire origin.

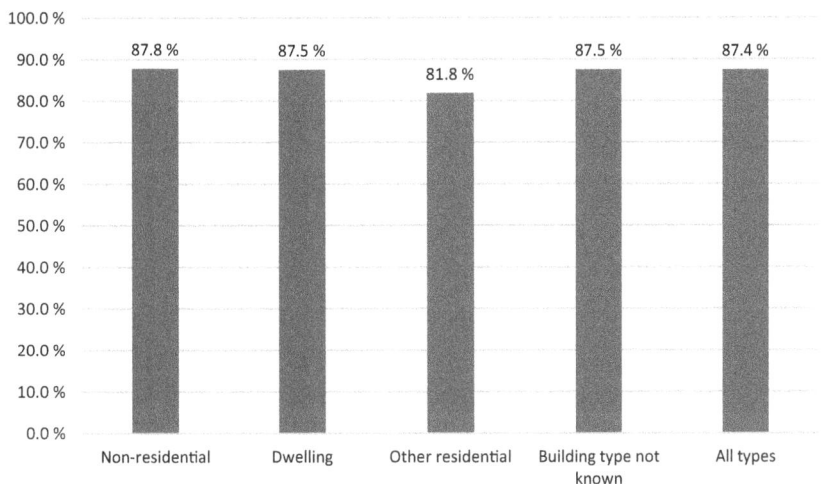

FIGURE 2.10
Operating sprinkler system in room of origin by building type.

Source: The OE report.

Two other categories are "On the same floor" and "Different floor." It is not clear whether this means there was a sprinkler in the room of origin, or whether there was a sprinkler, but it did not operate. If there was no sprinkler in the room of origin, why was there partial protection? This is not explained.

The OE report does not explain how a system can extinguish or control a fire up to 100% when it is not present in the room of origin, but rather is only on the same floor or even on a different floor. There is no explanation of why performance reliability improved from 532 cases, the basis for Figure 2.10. for known locations of the sprinkler system, to 677 cases in Figure 2.11. (Figure 8 in the report). Why can the performance data for the 532 cases not be used?

Under the assumption that sprinklers in the room of fire origin can extinguish or control the fire, as Figure 2.11. indicates, there is a possibility of determining the percentage of times when sprinklers operated effectively. The number of sprinkler fires in buildings was 2 268 from 2011 to 2015; the authors of the present work have adjusted this number to exclude the fires in 2016 (26).

The OE report does not explain why only 41% of the fires activated the sprinkler in the room. This is especially interesting, as 12.8% of the fires activated sprinklers on the same floor or a different floor, but not in the room

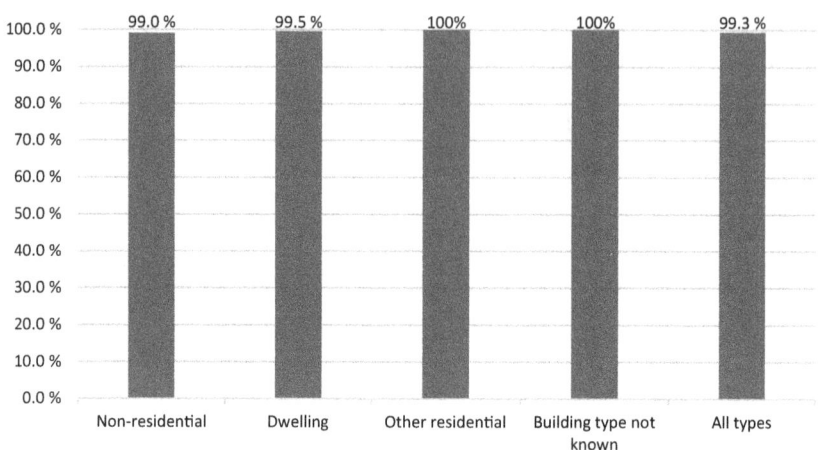

FIGURE 2.11
Impact on fires where system operated by building type.

Source: The OE report.

TABLE 2.20
Automatic sprinkler system reliability and effectiveness, by property use for 2011–2015 structure fires

A. All sprinklers (based on Fig. 1, 2, 5, 8 and Table 1, 4, 5 in the OE report)

Property use	Number of fires where sprinkler was present[*]	Per cent of fires not activating sprinkler[**]	Per cent (numbers) of fires per year	When equipment was present, fire was large enough to activate equipment, and sprinklers were present in fire area		
				Per cent where sprinkler operated in room of origin (A)[***]	Per cent effective of those that operated (B)	Per cent where sprinkler operated effectively (A x B)
Non-residential	1 705	65%	35% (603)	88%	99%	87%
Dwellings	409	33%	67% (273)	88%	100%	87%
Other residential	117	65%	35% (41)	82%	100%	82%
Not known	37	54%	46% (17)	88%	100%	88%
All fires	**2 268**	**59%**	**41% (934)**	**87%**	**99%**	**87%**

[*] These numbers are based on the total numbers of fires (2,294 minus 26 for 2016) and given in per cent.
[**] The OE report uses one decimal in the tables and none in the presentation of findings. The authors of this book do not use the decimal to be more in line with rest of the reports reviewed.
[***] The rest of the reports reviewed are only interested in systems present in the area of fire. A system that does not operate in the area of the origin fire cannot under any conditions be accepted as successful.

where the fire started. There is no explanation of how it was determined that there was insufficient heat to activate the sprinklers.

The OE report rates performance effectiveness as 99.3% and operational reliability as 94.3%. This gives total reliability of 99.3 x 94.3 = 93.6%. This contrasts with the finding of 87% in Table 2.20.

The OE report has data on the number of sprinklers activated for 788 of the 945 fires.

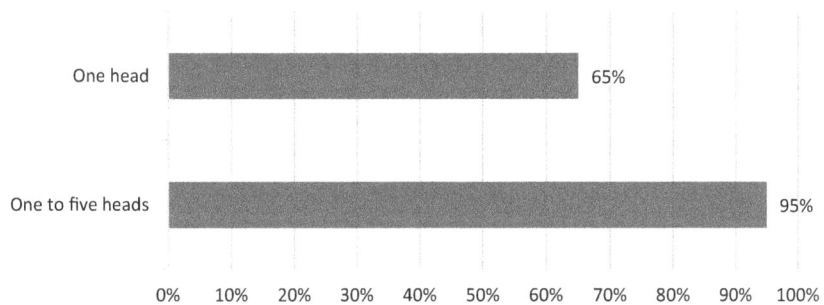

FIGURE 2.12
Per cent of fires where one or one to five sprinklers were activated.

Source: The OE report.

The 2017 NFPA report (National Fire Protection Association Research, 2017b) cites the per cent for one and one to five sprinklers as 79% and 97%, respectively. It concludes that reliability is 88%. Since it has a higher percentage for one and five or fewer sprinklers, this suggests the finding in Table 2.20. (i.e., the OE report's Figures 8 and 13) is more accurate.

Some of the possible answers in the IRIS Help and Guidance (DCLG, 2012) are unclear. For example, Question 7.15 asks "How many sprinklers operated? Select the number of sprinkler heads that operated." This question only applies to sprinkler systems. The person filling in the form can select 1, 2, 3, 4, 5, more than 5 heads, or "Not known." Why is "Not known" on this list, when it does not provide value to the data/report?

2.6.2. Unreliability

According to Figure 9 in the OE report, in 62% of all fires where a sprinkler did not operate, sprinkler systems were located in the room of the fire origin. This means sprinkler systems were outside the room 38% of the time. But according to Figure 10 in the report, 42.1% of the fires were in areas not covered by a sprinkler system. There is no explanation of this discrepancy.

According to the OE report (p. 7), the system operated in 945 cases and did not operate in 1 316 cases (p. 13). This represents 945 + 1 316 = 2 261 total cases. According to the report, in 2 294 cases, a sprinkler system was present. There is no explanation of why the other 33 cases are not included in the two subcategories; perhaps this is just an incorrect summary.

TABLE 2.21
Reasons for combined sprinkler failure and ineffectiveness: 2011–2015

Reason	Failure	
	Number	Per cent
Fault in the system	12	26%
System not set up properly	4	9%
System damaged by fire	7	15%
System turned off	18	39%
Operating failure	3	7%
Human error	1	2%
Flash fires	1	2%
Total	46	100%

2.6.3. Summary

The UK study does not provide definitions, and the assumptions are not explained to the reader. In addition, there are calculation errors, and there are no explanations of the differences in numbers from one case to another. Finally, the selection of cases seems random. However, this is the only report to use the word "indicates" about its findings.

2.7. Studies in Norway

This section of the chapter looks at a study done in Norway. It is included in the book to provide some data for Norwegian readers and explain how things have been done there.

2.7.1. Norwegian Experience With Sprinklers

Few studies have been done in Norway. Most look at the literature from around the world. One example is a study by Opstad (2002), using work by Bukowski (1999) as the main source for reliability data.

   One report that looks at raw data is the "Reliability of Automatic Sprinkler Systems—an Analysis of Available Statistics" (Malm, 2008). Statistics from Sweden, Finland, Norway, England, New Zealand, Australia, and the US are used in Malm's report. About 1.5 pages out of 52 are on Norway. Incident statistics come from *Direktoratet for samfunnssikkerhet og beredskap* (DSB) (Directorate for Civil Protection and Emergency Planning) from 1998 to 2007.

2.7.2. Reliability

Malm's report has the following definition of reliability: "Reliability refers to the probability that a sprinkler system will perform as expected. Reliability is the product of operational reliability and performance reliability." To this he adds, "Perform as expected expresses that a sprinkler system activates and contains, controls or extinguishes a fire" (p. 14). The origin of the definitions and the difference between contain and control are not clear.

A common mathematical expression for reliability is given as:

$$Reliability = Operational\ reliability \times Performance\ reliability$$

Malm's study for Norway uses the following mathematical expression without explanation:

$$Reliability = \frac{Number\ of\ incidents\ that\ functioned}{Total\ number\ of\ incidents}$$

The incident reports identify 1 262 fires in buildings with an extinguishing system. Out of these, 732 are identified as fires with sprinkler systems and the effect of the sprinkler systems is given for 453 cases. Four cases, including fire in boats, are removed.

TABLE 2.22
Automatic sprinkler system reliability by property use for 1998–2007 structure fires in Norway

All sprinklers			When sprinkler was present, effect known, and sprinklers operated effectively		
Property use	Number of fires per year where extinguishing equipment was present	Per cent (number) of fires with known sprinkler system	Per cent (number) of fires with known effect (A)	Number of fires where sprinkler operated effectively (A x B)	Per cent where sprinkler operated effectively (A x B)
All structures	1 258	58% (732)	36% (453)	334	74%

Source: The Malm report.

The report states that only 58% of fires with extinguishing systems were known to be sprinkler systems. The report does not explain if there was contact with fire departments to clarify the type of extinguishing systems, if they functioned, or the outcome.

There is no explanation of whether the extinguishing system was listed or whether the fire was too small or big enough to activate the system.

The raw data received from DSB raise even more questions about the work.

1. What value do the collected data have, when for over 25% of the registered fires, it is not known whether there was an extinguishing system present or not?
2. What value do the collected data have, when the data indicate that there were sprinklers but not if they functioned?
3. The report says, "The effect of the sprinkler systems is stated in 457 of the 736 incidents." In the data, all incidents are given as having one cause. What value does this bring to the report?
4. In the incident reports, the fire brigade must indicate "What stopped the fire spread." If a sprinkler system performs as expected to contain/control the fire, how is this analyzed and incorporated in the report if it is only possible to list one cause for stopping the fire spread?

2.7.3. Unreliability

The cause of failure is only known in 17 of the 118 incidents. These are shown in Table 2.23.

TABLE 2.23
Reasons for combined sprinkler failure and ineffectiveness: 1998–2007

Reason	Failure	
	Number	Per Cent
Not activated	8	47%
Out of order	6	35%
Insufficient amount of water	2	12%
Building partially sprinklered with deficient fire compartmentation	1	6%
Total	17	100%

Source: The Malm report.

The meanings are unclear. Does "Not activated" mean "System shut off," "Fire is large enough to activate sprinkler, but did not," or something else?

What value does "Not activated" give to the report if we cannot tell whether the sprinkler system should have been activated, or did not activate because the fire was too small?

2.7.4. Summary

Malm's study has a definition for reliability that is not used for the section on Norway. Instead, another expression is used.

The use of raw data is problematic, since Malm's study does not deal with uncertainty in the data.

2.8. What Is Known After the Review?

The preceding critical review has left many questions unanswered. It is not possible to find one given/accepted methodology to collect, analyze, and present reliability data. There is a lack of consensus in the fire engineering community, among various authorities, and across different insurance companies about the content and meaning of reliability, with the use of different words for reliability—like success, performance, operating, effectiveness, or operational effectiveness—for sprinkler systems.

Does this lack of consensus also indicate that data collecting, analysis, and presentation are done differently? If historical data on sprinklers are used as a basis to determine reliability, this must be done in a scientific way. Has this been done? The next chapter considers these issues.

3

Investigation of How Data Are Collected, Analyzed, and Presented in Selected Studies

Given the problematic definitions of reliability reviewed in the preceding chapter, could there be some more fundamental issues at stake, and, if so, how should they be reviewed?

The social sciences, including history, social anthropology, political science, socioeconomics, psychology, and so on, have certain methodologies. For the purposes of this book, two methodologies are especially relevant: how to conduct studies and how to analyze documents (Jacobsen, 2015).

3.1. Document Analysis

Document analysis, or source examination, is the analysis of documents (secondary data) to answer the research question (problem) by collecting and analyzing other words, phrases, stories on a topic, and reports. While a literature review tries to find theories or practices hole (or abundance), document examination is a systematic tool to examine all types of documents to find the answer to the initial question(s). This is helpful when:

1. It is impossible to get primary data;
2. A researcher wishes to learn how others have interpreted a situation, event, or data; or
3. A researcher wishes to learn what has been done or said.

Many different words are used to describe reliability (success, performance, performance effectiveness, operating, operating reliability, effectivity, operational effectively), and there are large gaps between the different levels or scores of reliabilities from one study to another. In addition, key findings are not always explained in a satisfactory way.

Thus, the question is: Is a study done in a satisfactory, scientific way? Are the results trustworthy? Can they be applied as documented expected reliability within the area of the study?

According to Jacobsen (2015), good scientific studies require the following steps:

1. Development of problem and purpose
   1.1. Is the issue clear?
   1.2. Is it descriptive or explanatory (causal)?
   1.3. Can it be generalized?

2. Choice of design
   2.1. Intensive (deep) or extensive (width) study design
   2.2. Descriptive or explanatory

3. Type of data (qualitative or quantitative)
4. Method of data collection
   4.1. Operationalization: how to make a concept measurable
   4.2. Design of the study
   4.3. Sources and use of sources

5. Selection and limitation of data
6. Analysis of data
7. Quality assurance of the analysis
   7.1. Conceptual validity
   7.2. Validation of contexts
   7.3. External validity
   7.4. Are the results trustworthy?

8. Discussion and presentation of results
   8.1. Methodological discussion
   8.2. Substantial discussion (connection of findings and theory)
   8.3. Presentation (also uncertainty)

This list is a systematic tool for the analysis of reports or studies of interest. Before the authors can look at how good these methods are in the field of fire science, the methods must prove themselves as they are, before any changes or adjustments are made.

3.2.   Document Analysis Validation

This book's authors suggest a systematic method of validation is to create a table based on the research phases given in the previous section. Validation using the questions in the eight steps given earlier is given in Table 3.1. by "Yes" (the question or step is done), "No" (the question or step is clearly not answered), or "Not sure" (it is not clear if the question or step is answered).

Table 3.2. gives a general overview of the steps that should appear in a scientific study.

The content of any secondary source is validated according to the template in Table 3.3.

The systematic tool of document analysis can now be used on the reports and studies of interest.

TABLE 3.1
Systematic overview of document analysis validation

Preparation and collection	Analysis	Presentation

1. Development of problem and purpose
  1.1. Is the issue clear?
  1.2. Is it explanatory (causal) or descriptive?
  1.3. Can it be generalized?
2. Choice of design
  2.1. Intensive (deep) or extensive (width) study design
  2.2. Descriptive or explanatory
3. Type of data (qualitative or quantitative)
4. Method of data collection
  4.1. Operationalization: how to make a concept measurable
  4.2. Design of the study
  c) Sources and use of sources
  4.3. Selection and limitation
	6. Analysis
	
	7. Quality assurance of the analysis
	  7.1. Conceptual validity
	  7.2. Validation of contexts
	  7.3. External validity
	  7.4. Are the results trustworthy?
		8. Discussion and presentation
		  8.1. Methodological discussion
		  8.2. Substantial discussion (connection of findings and theory)
		  8.3. Presentation (also uncertainty)

TABLE 3.2
General overview of document analysis validation

Preparation and collection	Analysis	Presentation
1. Development of problem and purpose 2. Choice of design 3. Type of data 4. Method of data collection 5. Selection and limitation	6. Analysis 7. Quality assurance of the analysis	8. Discussion and presentation

TABLE 3.3
Quality assurance of the steps in the document analysis

Main step	Sub step	Explanation
		Preparation and collection
	1.1. Is the issue clear or not?	If the purpose of the study is not clear, this means the purpose has not been revised over time, or adjusted according to available sources and data.
	1.2. Is it descriptive or explanatory (causal)?	Is the study more interested in how (descriptive) or why (explanatory)?
	1.3. Is it desirable to generalize the findings?	If it is desirable to generalize the findings, this pushes the benchmark up.
1. Development of problem and purpose		**Is there an understandable problem and purpose?**
	2.1. Intensive (deep) or extensive (width) study design	Extensive designs have many units in the study, but few variables. Intensive designs have many variables, but few units.
	2.2. Descriptive or explanatory study	The extent of the study increases exponentially when it goes from descriptive to explanatory: 1. Correlation of cause and presumed effect. 2. Cause precedes effect in time. 3. Control of all other relevant factors.
2. Choice of overall study design		**Is the overall study design understandable?**
3. Type of data (qualitative or quantitative)		**Can be of interest to have this as a separate point in some studies (e.g. social sciences), but not in this type of study. There is a need for quantitative answers.**
	4.1. Operationalization, how to make a concept measurable	Make abstract concepts, like reliability, operation, function, effect, and so on, into something measurable.
	4.2. Design of the study	Does the study use its own design or another study's design?
	4.3. Source and use of sources	Is it possible to use or change data collecting or design the study to accommodate fire brigades, or must it be independent?

Main step	Sub step	Explanation
		Preparation and collection
4. How data are collected		**Does the study have a workable, detailed design?**
5. Selection and limitation		**Selecting more than one system leads to more than one study. The study must limit all types of events that don't control over other relevant factors: for example, ships vs. buildings or fully vs. partially protected.**
Analysis		
6. Analysis		**It is very important to do the analysis using scientific methods, including the use of results.**
	7.1. Conceptual validity	After concretization of the concepts, it is very important to ask: Do the indicators measure what is of interest?
	7.2. Validation of correlations	If an explanatory (causal) problem/design is chosen, this puts a strong demand on the study. Is the question answered by the analysis?
	7.3. External validity	When there is a wish to generalize, it is important to ask if the analysis is wide enough and representative.
	7.4. Are the results trustworthy?	Can the way the study has been done be the reason for the result? Consider the level, time, and causality.
7. Quality assurance of the analysis		**How good are the conclusions drawn from the analysis?**
Presentation		
	8.1. Methodological discussion	Methodological discussion goes through the steps for quality assurance of the results, step 7 of this section, and examines how the study was conducted.
	8.2. Substantial discussion (connection of findings and theory)	Is this empirically consistent or inconsistent with other studies in this field? What are the connections between the findings and how they should be theoretically understood?
	8.3. Presentation (also uncertainty)	Is it transparent, logical, and readable?
8. Discussion and presentation		**How good is the presentation of conclusions?**

3.2.1. Studies in Australia and New Zealand

Table 3.4. shows the authors' findings for the validation of Marryatt's *Fire—A Century of Automatic Sprinkler Protection in Australia and New Zealand—1886–1986.*

Reference	Success, individually and average (%)	Applied area/focus/ comments	Comments
Marryatt (Marryatt, Rev. 1988)	95.3–100 99.5	Inspection, testing, and maintenance exceeded normal expectations, and higher pressures.	Data from 1886–1986.

TABLE 3.4

Document analysis of *Fire—A Century of Automatic Sprinkler Protection in Australia and New Zealand—1886–1986*

Preparation and collection		Analysis		Presentation	
1. Development of problem and purpose	No[4] No[1]	6. Analyzing	No[14]	8. Discussion and presentation	No[23] No[20]
1.1. Is the issue clear?	No[2]			8.1. Methodological discussion	No[21]
1.2. Is it explanatory (causal) or descriptive?	Yes[3]			8.2. Substantial discussion (connection of findings and theory)	No[22]
1.3. Can it be generalized?					
2. Choice of overall study design	Yes[7] No[5]	7. Quality assurance of the analysis	No[19] No[15]	8.3. Presentation (also uncertainty)	
2.1. Intensive (deep) or extensive (width) study design.	Yes[6]	7.1. Conceptual validity	No[16]		
2.2. Descriptive or explanatory		7.2. Validation of correlations	Yes[17] No[18]		
3. Type of data (qualitative or quantitative)	Yes[8]	7.3. External validity			
4. Method of data collection	No[12]	7.4. Are the results trustworthy?			
4.1. Operationalization: make a concept measurable	Yes[9] No[10]				
4.2. Design of the study	No[11]				
4.3. Source and use of sources					
5. Selection and limitation	No[13]				

[1] It is not clear what satisfactory performance is. Even if there is a definition, this is not in accordance with the statistics used in the book. Definition: "Fires which have either been completely extinguished, or controlled by the automatic sprinkler system to the point that they would by extinguished even if supplementary action had not been taken by fires brigades or others" (p. 18). Table 4 in Marryatt on overall performance redefines: "Fires which were completely extinguished by action of automatic sprinklers, fires in which hand extinguishers were used, and fires in which there was Fire Brigade action."

[2] The study is both.

³ The first line is "of this book, which covered the experience with automatic sprinkler system in Australia and New Zealand" (p. 13). This is a generalization.

⁴ The sum of 1.1., 1.2., and 1.3.

⁵ Both. Extensive when using around 9 000 fires with working sprinklers in this study and intensive when there are many variables.

⁶ Explanatory. Not only not controlled, but also causes/reasons (see Chapter 18 in Marryatt).

⁷ The sum of 2.1. and 2.2.

⁸ Quantitative.

⁹ The book is attempting to measure either sprinkler extinguished fire or fire on its way to being extinguished by the sprinkler system without interference. Even if this is not the case, this is not considered here. See footnote 10.

¹⁰ Study design does not measure this, but interference includes hand extinguishers and fires in which there was fire brigade action.

¹¹ It is not clear if Wormald International Group of Companies is the only supplier of records or only for some years. The source is not presented in a scientific manner.

¹² The sum of 4.1., 4.2., and 4.3.

¹³ It is not clear if partially protected buildings are included, why four fires in marine automatic sprinkler system are included, and why fires with water supplies shut off are taken out.

¹⁴ The report has several incorrect calculations, including summation of the number of events.

¹⁵ There is no discussion or validation of how results answer the question.

¹⁶ There is no discussion of correlation, i.e., that cause comes before effect in time and controls other conditions.

¹⁷ Only compared to NFPA.

¹⁸ The results are so different from other studies that the lack of proper discussion and incorrect calculations make the results appear untrustworthy.

¹⁹ The sum of 7.1., 7.2., and 7.3.

²⁰ Chapter 21 only touches on this subject.

²¹ Either fire theory (including extinguishing) or reliability theory is discussed or defined.

²² Lack of uncertainty analysis.

²³ The sum of 8.1., 8.2., and 8.3.

3.2.2. Studies in the United States

In this book, the authors have discussed three US studies: National Fire Protection Association (1970), National Fire Protection Association Research

TABLE 3.5
Overview of US studies of interest for document analysis

Reference	Success, individually and average (%)	Applied area/focus/ comments	Comments
1970 NFPA (National Fire Protection Association, 1970)	79.2–98.2 96.2	Data from 1897–1969 were 95.8% on average.	Data from 1897–1924 and 1925–1969.
2010 NFPA (National Fire Protection Association Research, 2010)	80–94 91	This study was done on sprinklers and other automatic fire extinguishing equipment.	Data from NFIRS 2004–2008.
2017 NFPA (National Fire Protection Association Research, 2017)	81–91 88	This study was done only for sprinklers.	Data from NFIRS 2010–2014.

(2010), and National Fire Protection Association Research (2017b). Since they are from same author, they are of special interest because they can more easily be compared.

3.2.2.1. US Experience With Sprinklers: 1970

The first study by National Fire Protection Association (1970) was published in *Fire Journal*, "Automatic Sprinkler Performance Tables, 1970 Edition." Table 3.6. shows the document analysis of the 1970 NFPA report.

TABLE 3.6
Document analysis of "Automatic Sprinkler Performance Tables, 1970 Edition"

Preparation and collection		Analysis		Presentation	
1. Development of problem and purpose	No[4]	6. Analysis	Yes[14]	8. Discussion and presentation	No[19]
1.1. Is the issue clear?	No[1]			8.1. Methodological discussion	No[16]
1.2. Is it explanatory (causal) or descriptive?	No[2]			8.2. Substantial discussion (connection of findings and theory)	No[17]
1.3. Can it be generalized?	Yes[3]				No[18]
2. Choice of design	Yes[7]	7. Quality assurance of the analysis	Not sure[15]		
2.1. Intensive (deep) or extensive (width) study design.	No[5]	7.1. Conceptual validity	Not sure	8.3. Presentation (also uncertainty)	
2.2. Descriptive or explanatory	Yes[6]	7.2. Validation of contexts	Not sure		
		7.3. External validity	Not sure		
3. Type of data (qualitative or quantitative)	Yes[8]	7.4. Are the results trustworthy?			
4. Method of data collection	No[12]				
4.1. Operationalization: make a concept measurable	No[9]				
	No[10]				
4.2. Design of the study	Yes[11]				
4.3. Source and use of sources					
5. Selection and limitation	Yes[13]				

[1] It is not clear what the following mean: equipment reliability/sprinkler performance/fire large enough/too small/operated effectively. There are no definitions or references to where they can be found.

[2] The study is both.

[3] The first line is "an over-all record" (p. 35). This is an attempt to generalize.

[4] The sum of 1.1., 1.2., and 1.3.

[5] Both. Extensive when using the NFPA Fire Record Department and over 75 000 fires and intensive when having many variables.

[6] Explanatory. Not only low performance, but also causes/reasons (Table 3 in the report).

[7] The sum of 2.1. and 2.2.

[8] Quantitative.

[9] Even if performance/effectiveness/reliability could be objectively measured, it is how the person filling in the form perceives performance/effectiveness/reliability that is noted and used in the analysis. There are no written instructions on how this validation/considering should be done, and there are no definitions; therefore, this could not be done the same way by everyone.

[10] Study design does not consider type of system.

[11] Insurance company and inspection bureau (must be considered impartial).

[12] The sum of 4.1., 4.2., and 4.3.

[13] Only buildings, but not system. Given a Yes under doubt.

[14] The report. Given a Yes under doubt.

[15] The sum of 7.1., 7.2., and 7.3.

[16] No.

[17] No.

[18] There is only positive uncertainty. A great number of small fires are not reported, and if they were included, this would have given higher performance.

[19] The sum of 8.1., 8.2., and 8.3.

3.2.2.2. US Experience With Sprinklers: 2010

The second study is from the National Fire Protection Association Research (2010), NFPA's Fire Analysis and Research Division, and is called "U.S. Experience with Sprinkler and Other Automatic Fire Extinguishing Equipment." The document analysis for the 2010 NFPA report is shown in Table 3.7.

TABLE 3.7

Document analysis of "U.S. Experience with Sprinklers and Other Automatic Fire Extinguishing Equipment," 2010

Preparation and collection		Analysis		Presentation	
1. Development of problem and purpose	No[4] No[1]	6. Analysis	Yes[14]	8. Discussion and presentation	No[21] Yes[18]
1.1. Is the issue clear?	No[2]			8.1. Methodological discussion	No[19]
1.2. Is it explanatory (causal) or descriptive?	Yes[3]			8.2. Substantial discussion	No[20]
1.3. Can it be generalized?				(connection of findings and theory)	
2. Choice of design	Yes[7]	7. Quality assurance of the analysis	No[17]		
2.1. Intensive (deep) or extensive (width) study design.	No[5]	7.1. Conceptual validity	No[15] Yes[16]		
2.2. Descriptive or explanatory	Yes[6]	7.2. Validation of contexts	Not sure	8.3. Presentation (also uncertainty)	
3. Type of data (qualitative or quantitative)	Yes[8]	7.3. External validity	Not sure		
4. Method of data collection	No[12]	7.4. Are the results trustworthy?			
4.1. Operationalization: make a concept measurable	No[9]				
4.2. Design of the study	No[10]				
4.3. Source and use of sources	Yes[11]				
5. Selection and limitation	Yes[13]				

[1] It is not clear what sprinkler reliability and effectiveness are. Even if there is no definition, there is a statement that effectiveness should be measured relative to design objectives, and the design purpose is to confine a fire to the room of origin. This has not been proved to be right.

(Continued)

TABLE 3.7 *(Continued)*

[2] It studies both.

[3] The first line is "Automatic sprinklers are a highly effective and reliable elements" (p. i). This is an attempt to generalize.

[4] The sum of 1.1., 1.2., and 1.3.

[5] Both. Extensive when using the US Fire Administration's National Fire Incident Reporting System (NFIRS 5.0) corrected with NFPA Fire Record Department and over 57 000 fires, and intensive when there are many variables.

[6] Explanatory. Not only low ineffectiveness, but also causes/reasons (see Figures 11 to 13 in the report).

[7] The sum of 2.1. and 2.2.

[8] Quantitative.

[9] Even if performance/effectiveness/reliability could be objectively measured, it is how the person filling in the form perceives performance/effectiveness/reliability that is noted and used in the analysis. There are no written instructions on how this validation/considering should be done, and there are no definitions; therefore, this could not be done the same way by everyone.

[10] Design of study does not consider type of system.

[11] NFPA study uses NFIRS to scale the numbers, but NFIRS is a voluntary system.

[12] The sum of 4.1., 4.2., and 4.3.

[13] Only buildings and excluding partially protected buildings, but not systems. Given a Yes under doubt.

[14] The report.

[15] Even if 49% of the fires were too small to activate equipment, there is no discussion or validation of what "too small" means or how this affected performance/reliability.

[16] Tables 4 and 5 show correlation; cause comes before effect in time and controls other conditions. The fact that "Fires too small to activate" is not supported by either qualitative or quantitative data is not considered.

[17] The sum of 7.1., 7.2., and 7.3.

[18] Section 1 and Appendix A in the report.

[19] Either fire theory (including extinguishing) or reliability theory is discussed or defined.

[20] Appendix A has a discussion of uncertainty, but this is not specified in numbers. Lack of uncertainty analysis.

[21] The sum of 8.1., 8.2., and 8.3.

3.2.2.3. US Experience With Sprinklers: 2017

The third US study is the most recent from National Fire Protection Association Research (2017b), "U.S. Experience with Sprinklers 2010–2014." The document analysis of the 2017 NFPA report is shown in Table 3.8.

3.2.3. Studies in the United Kingdom

The next work is from the UK: *Efficiency and Effectiveness of Sprinkler Systems in the United Kingdom: An Analysis from Fire Service Data* (Optimal Economics, 2017). The document analysis of this study is shown in Table 3.9.

Reference	Success, individually and average (%)	Applied area/ focus/ comments	Comments
NFSM (Optimal Economics, 2017)	92–97.7 93.6	United Kingdom	2017

TABLE 3.8

Document analysis of "U.S. Experience with Sprinklers 2010–2014," 2017

Preparation and collection		Analysis		Presentation	
1. Development of problem and purpose	No[4] No[1]	6. Analysis	Yes[14]	8. Discussion and presentation	No[21] Yes[18]
1.1. Is the issue clear?	No[2]			8.1. Methodological discussion	No[19]
1.2. Is it explanatory (causal) or descriptive?	Yes[3]			8.2. Substantial discussion	No[20]
1.3. Can it be generalized?				(connection of	
2. Choice of design	Yes[7]	7. Quality assurance of the analysis	No[17] No[15] Yes[16]	findings and theory)	
2.1. Intensive (deep) or extensive (width) study design.	No[5]	7.1. Conceptual validity	Not sure	8.3. Presentation (also uncertainty)	
2.2. Descriptive or explanatory	Yes[6]	7.2. Validation of contexts	Not sure		
		7.3. External validity			
3. Type of data (qualitative or quantitative)	Yes[8]	7.4. Are the results trustworthy?			
4. Method of data collection	No[12]				
4.1. Operationalization: make a concept measurable	No[9]				
4.2. Design of the study	No[10]				
4.3. Source and use of sources	Yes[11]				
5. Selection and limitation	Yes[13]				

[1] It is not clear what sprinkler reliability and effectiveness are. Even if there is no definition, there is a statement that effectiveness should be measured relative to design objectives, and the design is to confine a fire to the room of origin. This has not been proven to be right.

[2] The study is both.

[3] The first line is "Sprinklers are highly effective and reliable part. . . " (Abstract). This is an attempt to generalize.

[4] The sum of 1.1., 1.2., and 1.3.

[5] Both. Extensive when using the US Fire Administration's National Fire Incident Reporting System (NFIRS 5.0) corrected with NFPA Fire Record Department and over 46 000 fires and intensive when having many variables.

[6] Explanatory. Not only low performance, but also causes/reasons (Tables 4 and 5 in the report).

[7] The sum of 2.1. and 2.2.

[8] Quantitative.

[9] Even if performance/reliability could be objectively measured, it is how the person filling in the form perceives performance/reliability that is noted and used in the analysis. There are no written instructions on how this validation/considering should be done and there are no definitions; therefore, this could not be done the same way by everyone.

[10] Study design does not consider type of system.

[11] NFPA study uses NFIRS, but NFIRS is a voluntary system.

[12] The sum of 4.1., 4.2., and 4.3.

[13] Only buildings and excluding partially protected buildings, but not systems. Given a Yes under doubt.

[14] The report.

[15] Even if 44% of the fires were to be too small to activate equipment, there is no discussion or validation of what "too small" means or how this affected performance/reliability.

[16] Tables 8 and 9 show correlation; cause comes before effect in time and controls other

(Continued)

TABLE 3.8 *(Continued)*

conditions. The fact that "Fires too small to activate" is not supported with either qualitative or quantitative data is not considered.

[17] The sum of 7.1., 7.2., and 7.3.

[18] Section 1 and Appendix A in the report.

[19] Either fire theory (including extinguishing) or reliability theory is discussed or defined.

[20] Appendix A has a discussion of uncertainty, but this is not specified in numbers. Lack of uncertainty analysis.

[21] The sum of 8.1., 8.2., and 8.3.

TABLE 3.9

Document analysis of *Efficiency and Effectiveness of Sprinkler Systems in the United Kingdom*, 2017

Preparation and collection		Analysis		Presentation	
1. Development of problem and purpose	No[4] No[1]	6. Analysis	Not sure[14]	8. Discussion and presentation	No[23] No[20]
1.1. Is the issue clear?	No[2]			8.1. Methodological discussion	No[21]
1.2. Is it explanatory (causal) or descriptive?	Yes[3]			8.2. Substantial discussion	No[22]
1.3. Can it be generalized?				(connection of findings and theory)	
2. Choice of overall study design	Yes[7] Yes[5]	7. Quality assurance of the analysis	No[19] No[15]		
2.1. Intensive (deep) or extensive (width) study design.	Yes[6]	7.1. Conceptual validity 7.2. Validation of correlations	No[16] No[17] No[18]	8.3. Presentation (also uncertainty)	
2.2. Descriptive or explanatory		7.3. External validity			
3. Type of data (qualitative or quantitative)	Yes[8]	7.4. Are the results trustworthy?			
4. How to collect data	No[12]				
4.1. Operationalization: make a concept measurable	No[9] No[10]				
4.2. Design of the study	Yes[11]				
4.3. Source and use of sources					
5. Selection and limitation	No[13]				

[1] Five key questions are in Section 2.2. but there are no definitions, and the questions conflict with the Summary.

[2] The study is both.

[3] Summary: "The aim of the analysis was to provide an authoritative assessment of the reliability and effectiveness of sprinkler systems in controlling and extinguishing fires and in preventing damage" (p. 1).

[4] The sum of 1.1., 1.2., and 1.3.

[5] Extensive design with many units (around 2 300 fires), but few variables (Figure 4 in the report).

[6] Explanatory. Not only not controlled, but also causes/reasons for not operating (Section 3.3 in the report).

[7] The sum of 2.1. and 2.2.

[8] Quantitative.

[9] Even if performance/reliability could be objectively measured, it is how the person filling in the form perceives it that is noted and used in the analysis. The survey was not done on site, and there is an option to answer "Not known" in the IRS; therefore, there is no way this could be done the same way by everyone.

[10] Design gives "Not known" to measurable questions.

[11] The reports come from 47 fire and rescue services across the UK.

[12] The sum of 4.1., 4.2., and 4.3.

[13] The reason for incorporating systems not present in room of fire origin or same floor is not clear, and there are different numbers of fires in each figure and table.

[14] It is not clear how the analysis was done. Only 2 261 cases are presented, when there should be 2 294, and the use of data from one question to another is not clear. There is no clarification on the validation of selected numbers.

[15] There is no discussion or validation of how results answer the question.

[16] There is no discussion of correlation, i.e., that cause comes before effect in time and controls other conditions.

[17] There is no discussion of fires with such low heat that the sprinkler does not activate; nor is there a discussion of sprinklers not present in area of fire (partial protection), or fires that are large enough to activate the sprinkler, but do not.

[18] Because the study includes invalid numbers of fires and systems that operate on different floors and outside the room of origin without discussing the reason for doing so, the results are untrustworthy.

[19] The sum of 7.1., 7.2., and 7.3.

[20] Section 3.4 only touches on this subject.

[21] Either fire theory (including extinguishing) or reliability theory is discussed or defined.

[22] Lack of uncertainty analysis. The fourth conclusion states: "Operational reliability measures the probability that a system will operate as designed when required" (p. 18). No definition, discussion, or presentation of the design or the individual results supports this statement.

[23] The sum of 8.1., 8.2., and 8.3.

3.2.4. Studies in Norway

Document analysis is useful when it is impossible to collect primary data. In this case, it was possible to do so; therefore, there is no reason for the analysis. The conclusion of the literature review on the report by Malm (2008) was that it cannot be taken as historically reliable and cannot predict the reliability of sprinkler systems in Norway.

3.2.5. Summary of the Document Analysis

The basic question in this section is whether the analyzed studies can be used in a scientific way as a general basis for predicting the reliability of sprinkler systems. It must be stressed that the authors are not attempting to judge the value of the studies. Table 3.10. gives an overview of the findings.

If a study does not define its terms or its definitions, what value does it have to the field of reliability? Even if several factors indicate the study, in general, is done in a scientific way, *the conclusion is that it cannot be used to document reliability or predict the probability that sprinkler systems will function as designed.*

TABLE 3.10

Overview of document analysis validation for the examined studies

Reference	1.	2.	3.	4.	5.	6.	7.	8.	SUM
Marryatt (Marryatt, Rev. 1988)	No	Yes	Yes	No	No	No	No	No	**No**
1970 NFPA (National Fire Protection Association, 1970)	No	Yes	Yes	No	Yes	Yes	Not sure	No	**No**
2010 NFPA (National Fire Protection Association Research, 2010)	No	Yes	Yes	No	Yes	Yes	No	No	**No**
2017 NFPA (National Fire Protection Association Research, 2017b)	No	Yes	Yes	No	Yes	Yes	No	No	**No**
NFSM (Optimal Economics, 2017)	No	Yes	Yes	No	No	Not sure	No	No	**No**

[1] Development of problem and purpose
[2] Choice of overall study design
[3] Type of data
[4] How to collect data
[5] Selection and limitation
[6] Analysis
[7] Quality assurance of the analysis
[8] Discussion and presentation

What can be said with scientific certainty about sprinklers based on the studies in question? Findings for the 2017 NFPA research study (National Fire Protection Association Research, 2017b) appear in Table 2.18. The conclusion here is that 56% of the fires were large enough to activate the sprinklers, 92% were activated, and 96% were effective. This gives a probability of 56 x 92 x 96 = **49%** that a sprinkler will perform efficiently when there is a structure fire.

The premises for a scientific study are discussed in Section 3.4. But exactly how scientific is the use of document analysis? How well does this method from the social sciences work when it is applied in the natural sciences, in this case, in Fire Safety and Fire Engineering? This is discussed at greater length in Chapter 5.

3.3. Discussion of Findings

Are sprinkler systems the same in the United States, the United Kingdom, Australia/New Zealand, and Norway? Are they engineered in the same way, are planning and building activities the same in their requirements for engineering and installation, and are the control and maintenance regimes after the system is put into operation the same? They are not, and this explains some of the differences in the reliability findings in the different countries, but the main finding of the document analysis is that all studies have problems in four areas: development of problem and purpose (including a definition of reliability), how to collect data, quality assurance of the analysis, and discussion and presentation.

3.3.1. Development of Problem and Purpose

The first problem area is the development of the problem and purpose. Only one study has a definition (Marryatt, Rev. 1988), but not of reliability. Simply stated, reliability is the ability to function as intended. More precisely, it is the characteristic or expression of the ability of a component or system to perform an intended function. This includes the lifetime probability distribution of failure, statistical life expectancy, expected number of failures per unit of time, a system or component's ability to function satisfactorily over time and the likelihood that it will work at a specific time (Aven, 2006). For a sprinkler system, "intended" means the likelihood of functioning as *designed*. This includes correct or proper design following a sprinkler standard. This point is missed in every report or study mentioned in this book. Only with clear definitions (in this case, the probability of functioning as designed) can the development of a problem and purpose be done in a scientific way.

3.3.2. How to Collect Data

The second problem area, how to collect data, derives directly from the first. What does the ability to function as designed mean for a sprinkler system? Well, that comes down to what kind of sprinkler system is of interest.

What do American standards say about this? NFPA 13D (National Fire Protection Association, 2010a), "Standard for the Installation of Sprinkler Systems in One- and Two-Family Dwellings and Manufactured Homes," says the following:

> 1.2.1. The purpose of this standard shall be to provide a sprinkler system that *aids* in the detection and control of residential fires and thus *provides improved* protection *against injury and life loss.*
>
> 1.2.2. A sprinkler system designed and installed in accordance with this standard shall be *expected to prevent flashover* (total involvement) in the room of fire origin, where sprinklered, and *to improve the chance for occupants to escape or be evacuated.* (*authors' highlight*)

With a 7–10-minute rate for water demand for up to two residential sprinklers in the largest room, the sprinkler system is not designed to confine a fire in the room or design area, but to help people to escape from a fire by preventing flashover during the time the system has water and extending the escape time.

The purposes of sprinklers are expanded in the NFPA standard for larger residential buildings. NFPA 13R (National Fire Protection Association, 2010b),

"Standard for the Installation of Sprinkler Systems in Residential Occupancies up to and Including Four Stories in Height," specifies the following:

> 1.2.1. The purpose of this standard shall be to provide a sprinkler system that *aids* in the detection and control of residential fires and thus *provides improved* protection *against injury, life loss, and property damage.*
>
> 1.2.2. A sprinkler system designed and installed in accordance with this standard shall be *expected to prevent flashover* (total involvement) in the room of fire origin, where sprinklered, and *to improve the chance for occupants to escape or be evacuated. (authors' highlight)*

With a 30-minute rate for water demand for up to four residential sprinklers in the biggest room, the purpose is extended to include the prevention of property damage. This does not mean confined fires, just improved protection against property damage.

Standard NFPA 13 (National Fire Protection Association, 2016), "Installation of Sprinkler Systems," more clearly specifies protection of property:

> 1.2.1. The purpose of this standard shall be to provide a *reasonable degree of protection for life and property* from fire through standardization of design, installation, and testing requirements for sprinkler systems, including private fire service mains, based on sound engineering principles, test data, and field experience. *(authors' highlight)*

Chapter 3 in NFPA 13 gives the following definitions of purposes and sprinkler types:

> 3.3.11. *Fire Control.* Limiting the size of a fire by distribution of water so as to decrease the heat release rate and pre-wet adjacent combustibles, while controlling ceiling gas temperatures to avoid structural damage.
>
> 3.3.12. *Fire Suppression.* Sharply reducing the heat release rate of a fire and preventing its regrowth by means of direct and sufficient application of water through the fire plume to the burning fuel surface.
>
> 3.6.4.1.* Control Mode Density/Area (CMDA) Sprinkler. A type of spray sprinkler intended to provide *fire control* in storage applications using the design density/area criteria described in this standard.
>
> 3.6.4.3.* Early Suppression Fast-Response (ESFR) Sprinkler. A type of fast-response sprinkler that has a thermal element with an RTI of 50 (metres-second)$^{1/2}$ or less and is listed for its capability to provide *fire suppression* of specific high-challenge fire hazards.
>
> 3.6.4.9. Residential Sprinkler. A type of fast-response sprinkler having a thermal element with an RTI of 50 (metres-second)$^{1/2}$ or less that has been specifically investigated for its ability to *enhance survivability in the room of fire origin*, and that is listed for use in the protection of dwelling units. *(authors' highlight)*

It should also be noted that residential sprinklers specified by NFPA 13 have the four most hydraulically demanding sprinklers, regardless of room size, and water density should be double the requirement for a 13D/13R system.

NFPA 13 even mentions the possibility of reducing the area of operation when using quick-response sprinklers and designing the sprinkler system to suit the room design method (see Chapter 11, National Fire Protection Association, 2016).

The NFPA 13 standard defines the purposes of sprinklers based on the sprinkler type. Given this lead, why do studies not try to determine the reliability of specific sprinkler designs? They are being designed, built, and inspected/maintained according to sprinkler standards, and it is not clear why there is no interest in monitoring, controlling, and adjusting these standards based on their reliability. Recall that the reliability of a sprinkler system is its ability to function as designed to the specific sprinkler standard.

What about activation and performance? These are a natural part of reliability, but both the meaning of the words and their use need to be discussed and defined. Are activation and operation the same?

Not only must the number of triggered sprinklers be counted (and analyzed) for the sprinkler system in question, but two further areas also need attention: quality assurance and presentation.

3.3.3. Quality Assurance of the Analysis

The third area of concern is the quality assurance of the analysis. If there is no clear and measurable purpose, it is hard to check whether the results give the answer to the question. The process of quality assurance can be illustrated with a quality assurance wheel (Figure 3.1.).

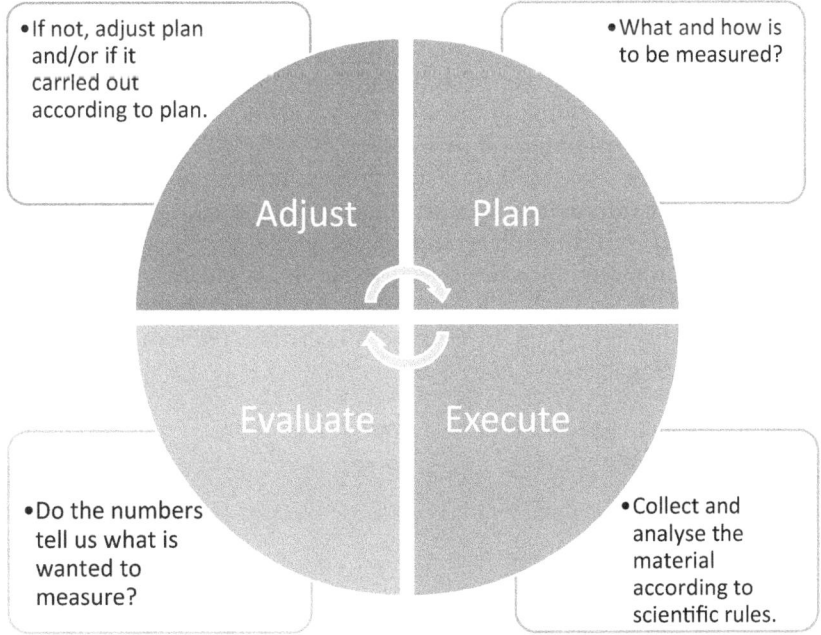

FIGURE 3.1
Quality assurance wheel.

This concept is valid for planning in general; it can also be applied to data used in studies of reliability and to the management of quality. It is very important to check results for conceptual validity, contextual validity, external validity, and trustworthiness. If they do not have these qualities, it will be hard to trust them.

3.3.4. Discussion and Presentation

The fourth problem area is presentation. Discussions, comparisons to earlier or other studies, trends, and honest views on uncertainty must indicate everything that could be a major issue. Only one study in the preceding literature review uses the word "indicates" when discussing the results (Optimal Economics, 2017).

3.3.5. Summary

The sprinkler has been used for 150 years with great success. Its ability to improve fire safety for both building occupants and fire and rescue personnel has been noted for a long time. The producers of sprinkler systems and insurance companies express great interest in testing and validating these systems' components. The same could be said about sprinkler standards. They too should be tested and validated with field experience. Although there are regulations in place, there is little follow-up on their usage and efficacy. Unfortunately, those making the decisions have not had sufficient interest or the knowledge required to conduct good studies. Knowing whether a mandatory regulatory demand is working as expected or not and collecting good quantitative data on performed-based design are of utmost importance. This data must be provided by the systems used by fire and rescue services, since they can give the impartiality required.

For the time being, the conclusion about the studies reviewed here is that they cannot be used to determine the reliability of sprinkler systems or the future probability of failure. Fire and/or sprinkler organizations do most of the work of producing studies that give some insight. Since the authors of this book are fire and reliability engineers, they find the lack of data on sprinkler reliability worrying. Even more concerning, there are few data on other fire protecting systems, both active and passive.

4

Reliability and Uncertainty Evaluations in Fire Sprinkler Systems

Section 1.5 in Chapter 1 notes that reliability is the characteristic of or the expression of the ability of a component or system to perform an intended function. Even a simple sprinkler system is likely to have an exponential growth in unreliability over time. If is it complicated (e.g. a system with pumps), start-up problems are likely to occur. The question is how long a sprinkler system will last and when the cost of maintenance or the possibility of failure justifies replacing it. The possibility of failure is often measured as mean time to failure (MTF).

4.1. Sprinkler Reliability

Intended function refers to the correct/proper design following a sprinkler standard. Sprinkler reliability is therefore the ability to function as designed to a standard for a particular fire hazard. As the authors point out in Section 3.3.2. of this book, there are minimal data on the reliability of different types of sprinkler systems. Data on a residential system expected to prevent flashover and to improve the chance for occupants to escape/evacuate are treated in the same way as an ESFR system that provides fire suppression; this clearly shows the need to understand reliability for different sprinkler types. Their performance is not the same; therefore, their reliability is not the same.

Why has this happened? The authors guess that the first sprinkler systems were not designed for a specific fire hazard. As the codes and standards came into place, research on how sprinklers performed in testing began, but there was an increasingly large diversity in types of sprinklers available. The rapid changes in quick response sprinklers, including residential ones that accelerated in the 1970s and 80s, illustrate this. When these changes occurred, few asked if they should have any consequences on how reliability was seen, examined, and treated.

Why has the lack of definitions discussed in Chapter 3 not been noticed and discussed, leading to changes in methodology? This is perhaps the most interesting question of all. If someone were to investigate this issue, the fire community would learn a great deal. This could lead to great changes in all fields of fire science.

For the time being, the fire community is waiting for someone to take on the task of collecting, analyzing, and presenting reliable data on how sprinklers operate and perform based on their function.

4.2. Operational Reliability/Operationality

Operational reliability, or operationality or operability, is a measure of the probability that a protection system or part of it will operate when needed, with testing and standards at a component level. It requires effort, time, money, development, and tests to get approval based on each component in a sprinkler system today. The Omega sprinkler failure (Fire Engineering, 1997) is probably the best-known example of what can happen when there are no demands to get a component approved. With testing, the knowledge of lifetime challenges has increased, leading to more demands for inspection and maintenance.

The lack of maintenance is an important reason for sprinkler failure in most studies mentioned in this book, except for Marryatt (Marryat, Rev. 1988). The question is: Should lack of maintenance or someone shutting off the main valve of the water supply be considered under operationality and reliability? Or are we interested in the reliability (operationality x efficiency) when the system is installed according to the right standards for a specific fire hazard (fire load), with approved components, and tested, inspected, and maintained according to the correct intervals? If the latter is true, the authors believe the fire community must agree on the terms and definitions of interest. Once this happens, the investigation of unreliability when, for example, valves are not supervised, maintenance is not done, and so on, can be done in an agreed-upon way.

The second challenge is robustness. Should we assume a system should have some sort of robustness, and should this be for the component or the system? One system that appears to have little robustness is the ESFR sprinkler system. It is well known that obstructions can have a huge impact on the system's suppression abilities. But does this have to do with incorporated operationality, or with efficiency, if the demands are obstacles and their placing? Where the robustness should be placed is not always clear. The same could be said for residential sprinklers. If the design area is two sprinklers, there is no room for failure; both sprinklers must work.

It is the authors' opinion that if there is different inherent robustness for each system, reliability will also differ.

4.3. Performance Reliability/Efficiency

Performance reliability or efficiency (performance is probably not the best word; efficiency is of more interest) is a measure of the adequacy of the system to successfully perform its intended function under specific fire scenario conditions.

Certain standards are for robustness, at least in some way. Some standards like the maritime IMO[1] A800 and the revised MSC.265(84) (International Maritime Organization, 2008) have fire tests with a disabled nozzle. This has interesting implications for the reliability of systems and how they perform in real time events. The most interesting part is that when there are no data based on their design under different standards and hazard classes, there is no way of knowing if the hazard class is correct.

The testing of a sprinkler happens with a designated fire load and geometry; its failure or success is often based on the same criteria, for example, that no more than four sprinklers are released and control the fire from spreading or the fire is put out. Once a database on standards and hazard classes is established, there will be evidence of robustness for some or all hazard classes and systems. It will also be possible to change the parameters in the standards.

4.4. Control of All Factors and Uncertainty

What factors contribute to the perception of what reliability is and what factors belong to system reliability? As discussed above, there are several perceptions of what a sprinkler system should do and therefore of what kind of reliability is of interest.

While some studies include partial protection (Optimal Economics, 2017), others are not clear if they do (Marryatt, Rev. 1988), and still others do not (National Fire Protection Association Research, 2017b). As a result, it is confusing to use and compare the studies. This stresses the importance of understanding fire dynamics, extinguishing theory, and the purpose of the used standard. If it is not within the standard to stop a fire that occurs on the outside of the building, this aspect should not be included in a study on reliability. It is more important to determine if laws and regulations were

---

[1] The International Maritime Organization is a specialized agency of the United Nations responsible for measures to improve the safety and security of international shipping and to prevent pollution from ships.

followed and suggest possible changes in practice and/or regulations. There is a big difference in how (and if) this is done in different countries.

Interestingly, even if uncertainty is a common part of life, this is seldom mentioned in the literature on reliability. A good example is Marryatt's book, discussed at length in the previous chapter (Marryat, Rev. 1988). As shown in Section 2.2. on Australia and New Zealand's experiences with sprinklers, probably all of Marryatt's facts came from a sprinkler manufacturer. In addition, these data had not been looked at with a critical scientific eye. This might lead us to suspect that a manufacturer's desire for positive results could influence the reports of the results. This could be dealt with in several ways, but the first step is openness about the source of information. When the impression is that this is concealed, it is not possible to create trust. Without trust, the work has no scientific meaning.

Why is it so difficult to find serious scientific intent in so much of this work? Why are people not asking good scientific critical questions? And when questions are asked, why are those responsible not willing to answer? It seems reasonable that the manufacturer should have a genuine interest to know how its sprinkler (and those of others) performs.

The information from the manufacturer should have been treated like all other sources. The facts must be verified. Marryatt could have collected all fire reports from one fire brigade, within a specific period, and then compared the reports from the manufacturer to the reports from fire brigade. How many fires does the fire brigade report in sprinkled buildings, and how many does the manufacturer mention? Is it the same? How much do they differ? What was the outcome of the reports? Did the manufacturer make any changes? Where there is failure, there must be further examination. Have cases of success, for example, when a fire was small and only one sprinkler activated, been reported to the manufacturer? Is there an underreporting of success, and if so, why? Does the fire brigade receive a fire alarm for all fires? When did the fire brigade start to collect reports on fires, and what facts have they collected?

These and other questions must be asked by researchers. When they know the answers, they can address uncertainty.

5

Develop Methodologies and Proposals for Studies With General Scientific Value

Most of the sprinkler standards in the world say something like: "The purpose of this standard shall be to provide a sprinkler system that aids in the detection and control of fires." This, or a similar statement, is written into the purpose of the standard or communicated throughout the standard. Since the beginning of sprinkler systems, the purpose of detecting and warning of fire has been a natural part of the system. Even today, a sprinkler control valve is called an alarm valve (see Figure 5.1.). Therefore, those who plan to conduct studies on sprinkler systems should consider detection and warning. If this is of no interest, what is the purpose of having this as part of the design and the physical installation of sprinkler systems?

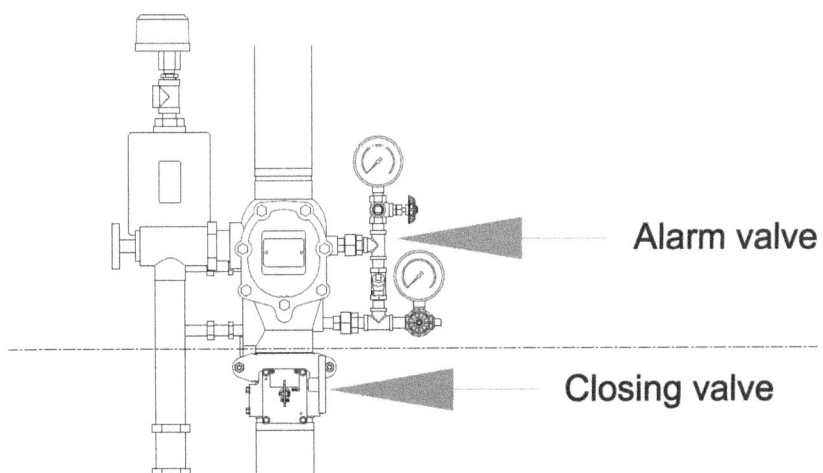

FIGURE 5.1
Sprinkler control valve with its two main components.

As mentioned in the previous chapter, certain areas in the field of Fire Engineering require improvement. This chapter should be of some help, as it offers suggestions for methodologies to collect, analyze, and present reliability data for all types of safety systems, with a focus on sprinklers.

5.1. Methodology

Many factors in how things are done affect both the quality of the data and the outcome of the analysis. One example is from the UK Incident Recording System (IRS) (DCLG, 2012). After the answer "Yes" is given to the question "Did the safety system operate?", the next question is "Select the number of sprinkler heads that operated." This can be answered by 1 to 5, more than 5, and "Not known." Why is "Not known" given as an option at all? Is it not assumed that visual inspection can give answer on this? Furthermore, will this type of question improve the data or not? These and other questions must be asked. Data collection and analysis is a multi-discipline and multi-team effort.

Based on the document analysis performed previously and the requirements of a scientific study already enumerated in this book, in Table 5.1., the authors suggest a revised format for a scientific study.

TABLE 5.1
Division and steps in a study of sprinkler reliability

Main step*	Sub step	Explanation
	Preparation and collecting	
	1.1. Is the issue clear or not?	If the purpose or problem of the study is not clear, this often means the purpose has not been revised over time, or adjusted according to available sources and data.
Informative	1.2. Is it descriptive or explanatory (causal)?	Is the study more interested in how (descriptive) or why (explanatory)?
Informative	1.3. Is it desirable to generalize the findings?	If it is desirable to generalize the findings, this pushes the benchmark up.
1. Development of problem and purpose		**Does it have an understandable problem or purpose?**
Informative	2.1. Intensive (deep), extensive (width) study design, or both?	Extensive designs have many units in the study, but few variables. Intensive designs have many variables, but few units. If both are selected, there will be many units and variables.
Informative	2.2. Descriptive or explanatory?	This is of most interest if an explanatory design is selected. The following steps must be taken, as the extent increases exponentially when the study goes from descriptive to explanatory. 1. Correlation of cause and presumed effect. 2. Cause must precede effect in time. 3. Control of all other relevant factors.

Main step*	Sub step	Explanation
2. Choice of overall study design		**Is there an understandable overall design?**
	3.1. Operationalization, how to make a concept measurable	Make abstract concepts, like reliability, operation, function, effect, and soon on, into something measurable.
	3.2. Design of the study	Does it use its own design or another study's design?
	3.3. Source and use of sources	Is it possible to use or change data collecting or design the study to accommodate fire brigades, or must it be independent?
	3.4. Selection and limitation**	The selection of more than one extinguishing system will result in more than one study. The study must limit all types of events that do not control other relevant factors: e.g. ships vs. buildings; fully vs. partially protected; residential vs. ordinary hazard; and so on.
3. How data are collected		**Does the study have a workable design?**
Analysis		
4. Analysis		**It is very important to do the analysis by scientific methods, including control of the use of results.**
	5.1. Conceptual validity	After concretization of the concepts, it is very important to ask: Do the indicators measure what is of interest?
	5.2. Validation of correlations	If an explanatory (causal) problem/design has been chosen, this places a strong demand on the study. Is it answered by the analysis?
	5.3. External validity	Is it possible to go from empirical evidence to known theory, based on the findings? If is desirable to generalize from selected units, it is important to ask if the study is wide enough and if there is a representative selection in the analysis.
	5.4. Are the results trustworthy?	Can the way the study has been done be the reason for the results? There is a need to control the level, time frame, and causality.

(Continued)

TABLE 5.1 *(Continued)*

Main step*	Sub step	Explanation
5. Quality assurance of the analysis		**How good are the conclusions drawn from the analysis?**
	Presentation	
	6.1. Methodological discussion	Methodological discussion includes the steps of quality assurance of the results (step 5 of this section) and how the study has been conducted.
	6.2. Substantial discussion— connection of findings and theory	Is it empirically consistent or inconsistent with other like or similar studies in this field? What are the connections between the findings and how this should be theoretically understood?
	6.3. Presentation (also uncertainty)	Is it transparent, logical, and readable?
6. Discussion and presentation		**How good is the presentation of the conclusions?**

* Former step 3 type of data (qualitative or quantitative) has been removed in this table, because qualitative data, analysis, and presentation are seldom of interest in Fire Safety.
** While this is step 5 in the original document analysis, several aspects argue for its inclusion in step 3. The need to select and limit the type of objects, systems, and so on when designing the study does not exclude the possibility of excluding collected data that are obviously wrong or inadequate later in the process.

Further comments on the suggested methodology:

1. Development of problem and purpose: It would be a major improvement if the international fire community could reach an agreement on the terms used in reliability data. There are many different words used to describe reliability or part of reliability.

   Of real interest is a study that wishes to generalize the outcome. It is important to find out if this is possible. Are the sources available for the whole area of interest? If not, perhaps an indicative study could be used to illuminate the problem or to address the proper authorities for improved data collection.

2. Choice of overall study design: Figure 5.2. shows the correlation between intensive and extensive study designs.

   An extensive study with many units and few variables is a very good foundation for a generalizable study.

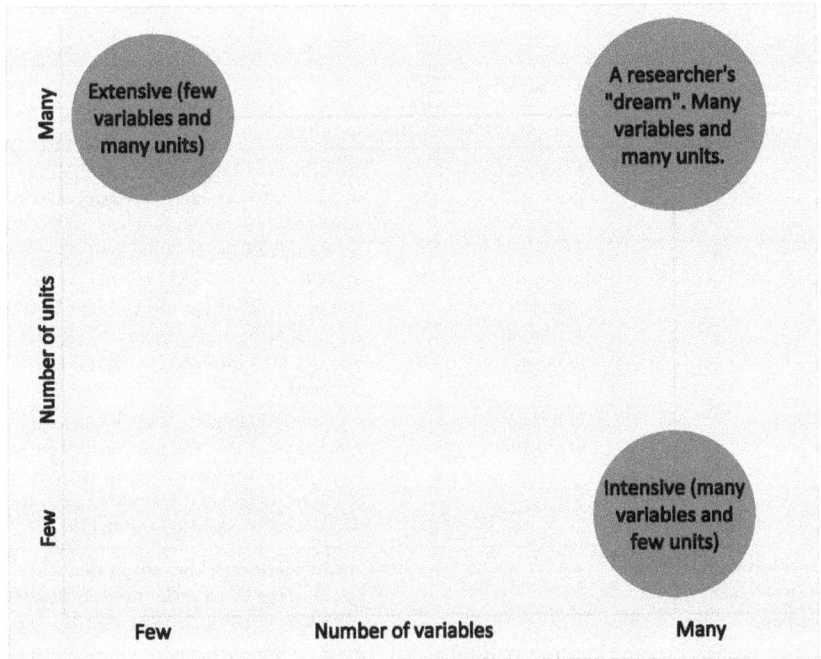

FIGURE 5.2
Classification of study design as intensive or extensive.

3. There is no right or wrong choice, but the choices must be conscious. Every choice has consequences for collecting, analyzing, and presenting.

5.2. How to Perform a Simple Study (Descriptive Study)

Based on the overview in Table 5.1., this section gives examples of two different kinds of studies, starting with the simplest.

5.2.1. Step One: Development of Problem and Purpose

The first step is to determine the overall design and what is of interest. This requires a detailed plan of what kind of sprinkler systems are of interest, and how, where, and over what time. Table 5.2. suggests how a simple descriptive study can be designed.

TABLE 5.2
Design of simple study

Main step	Sub step	Task
	Preparation and collecting	
	1.1. Is the issue clear or not?	The purpose of this study is to find the reliability (to function as designed) of sprinkler systems in Norway. Design means the selected sprinkler system.
	1.2. Is it descriptive or explanatory (causal)?	This is a descriptive study and causal reasons (why the systems work or do not work as intended) will not be covered.
	1.3. Is it desirable to generalize?	It is desirable to generalize historical reliability, as this is a good indicator of future probability of the sprinkler systems being designed and installed under the conditions covered by the study.
1. Development of problem and purpose		**Conduct a descriptive study that generalizes the national reliability of sprinkler systems in Norway.**
	2.1. Intensive (deep), extensive (width) or both study design	Based on the purpose, an extensive design with many units in the study and few variables is chosen.
	2.2. Descriptive or explanatory	Descriptive design is chosen.
2. Choice of overall study design		**Create an extensive and descriptive overall study design.**
	3.1. Operationalization, how to make a concept measurable	Definition: **Sprinkler system activation**: 1: the sprinkler control valve (alarm valve) opens; 2: the pump (if installed) starts; and 3: the sprinkler alarm activates. **Fire controlled by sprinkler system**; the fire is contained/extinguished within the sprinkler system's design (number of activated sprinklers and square metres covered).
	3.2. Design of the study	Based on design and event tree analysis, create a form for the study.
	3.3. Sources and use of sources	It is desirable to use the Norwegian BRIS (Brann, Redning, Innrapportering, Statistikk/Fire, Rescue, Reporting, Statistics)[1] for reporting fires.
	3.4. Selection and limitation	Only buildings fully protected by sprinklers or parts of buildings that are separated by fire section walls are of interest.

Main step	Sub step	Task
	Preparation and collecting	
		Only sprinkler systems by NS-EN 12845:2004 and CEA 4001 will be examined.
		Data on system type will always be validated against the ESS (Elektronisk System for Sprinkleranlegg/Electronic System for Sprinklers.[2]
		Fires from 2010–2014 will be examined.
3. How to collect data		**The design of the study is based on Norwegian BRIS, looking at building fires from 2010–2014 in buildings protected following NS-EN 12845:2004 and CEA4001 rules. Details on sprinkler systems and hazard classes are validated using ESS.**

[1] Norwegian incident reporting system used by all Norwegian fire and rescue personnel.
[2] ESS is the Norwegian insurance company's database for reports on inspections of existing sprinkler systems and for project reports on planned sprinkler systems.

Before determining how to collect data, for example, using earlier methods or taking data directly from BRIS, an event tree analysis of some nature must be performed to make sure the project description is correct, and every aspect is thought out to create tools for data collection and quality assurance.

5.2.2. Step Two: Choice of Overall Study Design

In the event tree shown in Figure 5.3., the starting point is: Fire in sprinkled building.

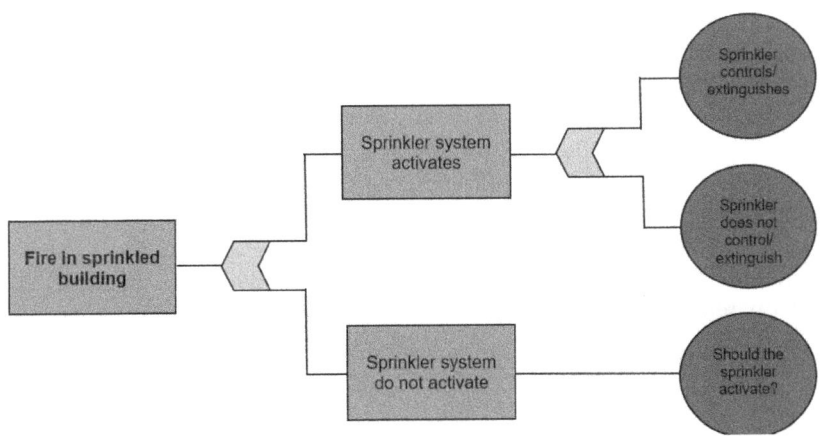

FIGURE 5.3
Event tree, simple design.

Before going further in the example, the three outcomes shown on the right side of the figure need discussion and clarification. There is also a need to discuss the use of the words activate/activation.

Activate/activation: In most of the literature, the words operate/operating/ operational are used about a system that starts to do something. This is interesting, but a system should operate *because* it was activated. If a system is activated, but does not operate, this has a cause. Even if the cause is not of interest in this case, the establishment of an accurate and precise vocabulary is necessary, and the word activate will be used herein.

Outcome 1: "Sprinkler system controls or extinguishes the fire." This outcome concerns the number of sprinklers activated and the area damaged by the flames, as the hazard classes have an area of design and a number of sprinklers that should be activated. The fire response team has no difficulty concluding that a fire has been put out or is under control, with only minor measures required to put out the fire. The number of activated sprinklers should relate to the area of flame damage. If not, this needs further investigation.

Outcome 2: "Sprinkler system does not control or extinguish the fire." This outcome concerns the number of sprinklers activated and the area damaged by the flames, for the same reason as Outcome 1 (control of hazard class and quality assurance), and whether the right outcome has been chosen. It is possible that the person filling in the report form thinks, for example, that 10 sprinklers and 100m² do not indicate control, but for every hazard class over OH1 in EN 12845, this is less than the design area for the sprinkler system. It is important to have control of all possible non-negative outcomes in the category of negative outcomes.

Outcome 3: "Sprinkler system does not activate." As shown earlier, this outcome is often neglected when it comes to quality assurance. It is very important to find out two things. One, was the system activated, but did not operate? Two, should it have activated but did not?

One: To determine this with a high degree of certainty, it is necessary to conduct an alarm test of the sprinkler valve. With an alarm, the system is intact. No alarm is a fault and indicates a problem with the system in general.

Two: The second question can be answered by a visual inspection of affected sprinklers. If the bulb/fusible link is intact and there is no suggestion that the heat was sufficient to activate the sprinkler, it can be concluded that the system should not have been activated. If there is uncertainty, this should be noted.

In the not-so-distant future, multi-sensor smoke detectors (including temperature reading) and digitalized control valves/pumps will give better quality data.

5.2.3. Step Three: How to Collect Data

Step three is writing the inquiry form to collect data. This is basic, regardless of how it will be distributed or how information will be collected from databases. The main purpose is to have control of questions that are of interest and to use them in the quality assurance of the analysis. Table 5.3. gives a sample inquiry form.

Even this simple form has eight questions about the consequences of the fire. By concretizing the activation/outcome in only yes/no answers, question 11 is changed from a qualitative question in the form's *Outcome 3* (consequences of the fire) to a quantitative question.

The most important concern is the quality of the reports from the fire brigade. If it is not possible to find the desired information (all or some)

TABLE 5.3
Inquiry form for fires in buildings protected by sprinkler systems

Form for fires in buildings protected by sprinkler systems	
Information about the building[*]	
1. Address:	Official identification number (Norway Gnr/Bnr)
Type of building:	Is the building registered in ESS? If not, must the owner must do this?
Date and time of fire:	
Installing year and latest inspection/ maintenance:	Data from documentation or from ESS.
Information about the sprinkler system	
2. Type of sprinkler system:	CEA 4001 NS-EN 12854 NS-INSTA 900 NFPA 13 Other (specify)
3. Hazard class	OH/HHP/HHS 1–4
Information about consequences of fire	
4. Did the sprinkler system activate alarm (by the fire alarm system or by external alarm bell)?	Yes/No
5. Did the sprinkler system activate sprinkler pump if present?	Yes/No/Not present
6. Did the sprinkler system control or extinguish the fire?	Yes/No
7. If yes, how many sprinklers were activated and what was the area of flame damage?	Number: Affected area: m^2
8. If no, how many sprinklers were activated and what was the area of flame damage?	Number: Affected area: m^2
9. If the fire did not activate the sprinkler system, is the bulb or fusible link on sprinkler destroyed/damaged?	Yes/No
10. If the fire did not activate the sprinkler system, does a test of the alarm show it works?	Yes/No
11. If the fire did not activate the sprinkler system, are there any indications that it should have been activated (size of fire, damage in the area, etc.)?	Yes/No Affected area: m^2

[*] It must be possible to trace the data to the source, if needed.

about sprinklers in the fire brigade's system BRIS, how can this be done? How should questions be distributed and collected? Is it possible to get fire brigades to change their form to include the required questions? This is perhaps not possible, or may not be desired by the authorities. What then? Is it possible to get the fire brigades to use a sprinkler questionnaire in addition to their own? If not, is it possible to be notified about fires in sprinkled buildings, so that researchers can check out the buildings themselves?

All these questions stress the importance that a study is no better than the team that conducts the study. With no or only partial interest by the authorities in reliable data, it will be hard to find good information.

5.2.4. Step Four: Analyzing and Calculating, Including Uncertainty

In this step is it important to identify the uncertainty first. If there is no control of uncertainty, the calculations cannot be done correctly.

The first area that stands out is information about the building in question. Is there information about the sprinkler system? Even if many sprinkler systems are registered in ESS, there are probably many that are not. What effect does the missing information have on a study? Is it possible to receive such information?

The second area is to compare the data collected by the fire brigade or in partnership with them. Are they consistent? Does the area of fire damage correspond with the number of sprinklers released? Does the perception of control relate to the actual event? If not, this is a strong indication that the incident has not been perceived correctly. Perhaps there was more control than previously thought?

If it is possible to get complete information from the national BRIS, it may be possible to get information from one or more of the fire brigades. With no knowledge of the work in this area, however, it may be difficult to figure out how to collect data of scientific value. As pointed out in this book, there is no reason to hope that the authorities will do it. Without a serious approach, most work will stop at this point. The use of personal connections may be a good option.

There must also be some numbers on how many fires are reported and how many are not. As the authors have personally experienced, small fires with only one sprinkler head released and putting out the fire are not always reported to the fire brigade. If the sprinkler and fire alarm were not connected to the fire brigade or another alarm company, and no one called the fire brigade, there will be no trace of this in BRIS. So, one of the basic questions that must answered is how many fires with a positive outcome, and with only one sprinkler released, are we talking about? Would it be a good idea to send out a questionnaire to sprinkler entrepreneurs to get some information from them?

In the calculations of the number of fires and sprinkler activations, whether the research team wishes to give the mean, median, mode, or just one of the average values is up to the team, but what kind of value we are talking about and the numbers the team uses must be traceable.

Calculations must be done in a normal matter, with no questions on the process, and the results must be consistent throughout the study.

5.2.5. Step Five: Quality Assurance of the Analysis

By assuring the quality of the analysis, the fundamental requirements for a solid presentation in the last step are mostly met. The importance of checking the answers given against the purpose of the study cannot be stressed enough. The list in Table 5.4. is a minimum layout; it can be extended.

TABLE 5.4
Design of simple study, step five

5.1. Conceptual validity	**Overall question: Do the indicators measure what is of interest?**
	1. How many building fires were there per year during 2010–2014?
	2. How many fires were in sprinkled buildings?
	3. How many sprinkler systems were CEA 4001/ NS-EN 12845/NS-INSTA 900/NFPA 13/Other system?[1]
	4. How many fires activated the sprinkler system (both overall and by alarm and pump)?
	5. How many fires were indicated as controlled/ extinguished by the report?
	6. How many sprinklers were activated on the different systems and hazard classes? This also includes non-controlled fires.
	7. When the system did not activate, was sprinkler bulb/fusible link destroyed/ damaged?
	8. Are there indicators that the system should have activated/worked?
	9. Did the alarm test work on the non-activated sprinkler system?
5.2. Validation of correlations[2]	1. How many fires were controlled/extinguished according to the number of sprinklers and area of damage?
	2. Is the area affected by the fire (flames/hot plume) correlated to the number of activated sprinklers?[3]
	3. Are the area and number of sprinklers activated for a non-controlled fire correlated to the stated effect?[3]
5.3. External validity[4]	1. Are some geographic areas or fire brigades excluded from this study?
	2. Do the findings support known theory?
	3. Is uncertainty quantified?

(Continued)

TABLE 5.4 *(Continued)*

	5.4. Are the results trustworthy?	Once again, it is important to ask: Is it the way the data are collected (e.g. the inquiry form) and analyzed that produces the result?
		One example of *level failings* is when the results show that most fires are confined to the room of origin, and the conclusion is that sprinklers are designed to confine fire to the room of origin. Do data support this, and is the design based on the standards and test protocols? When the conclusion takes the result up or down a level, these failings occur.
		One example of a *time frame failing* is assuming a sprinkler system is no less reliable because of its age. For example, a study may look at a selected sprinkler system five years after installation and conclude there is no difference in reliability based on age.
		Causality failings will be discussed later under the explanatory study, but these affect all studies. One example is the reason a sprinkler system does not activate. With no cause for this, the conclusion can be that it should not activate, but there are no data to support such a conclusion. There is only certainly/uncertainty.
5. Quality assurance of the analysis		**Based on validation of concept, correlations, external information, level failings, time frame failings, and causal failings, how good are the conclusions drawn from the analysis?**

[1] Even if the CEA 4001/NS-EN 12845 system is of interest, it is possible to get some data on all types of systems at this point. With data here, it is possible to get an idea of other types of systems in Norwegian buildings.

[2] Validating correlations must be done carefully for an explanatory (causal) study (this will be reviewed in more detail in the next example).

[3] When it comes to uncertainty, this can be handled strictly theoretically by only looking at the collected data, or more practically by investigating. Case studies give more insight into how fires and sprinklers are connected and generate more correct uncertainty data.

[4] When the study generalizes from collected data to a further probability, some areas need validation.

As a result of a quality assurance check, the form may need to be corrected, and the analysis redone in some areas, but this is to be expected. In some instances, this will indicate areas not previously thought of, but the data are collected and cannot be changed in an economical/practical way. This suggests the need for comments on uncertainty to strengthen the sense that the study has been conducted in a scientific way.

5.2.6. Step Six: Discussion and Presentation

The last step is to finish the report and present the results of the study. The areas listed in Table 5.5. must be not be forgotten.

TABLE 5.5
Design of simple study, step six

Presentation		
	6.1. Methodological discussion	A theoretical review of the methodology of the study design, including data collection, analysis, and quality assurance of the results. 1. How good is the reliability of the data? 2. How good is conceptual validity? 3. How good is the internal validity? 4. How good is the external validity? These four questions constitute the total validity of the study.
	6.2. Substantial discussion— connection of findings and theory	*Findings*: Are the results consistent or inconsistent with earlier and other similar studies? When there are earlier studies, what are the long-term changes? *Theory*: What is the connection between results and known theory? What is unknown? Can some hypotheses be followed up on in later or other studies? *N.B.*: What do the findings tell us about the standards used? What are the suggestions for *further work*?
	6.3. Presentation (also uncertainty)	There must be a definition list or presentation of key terms earlier in the report. The summary or abstract of findings helps the reader understand the report. With many options on how figures and tables can be designed, there is no need to use colours and shapes that take attention away from the facts. Make the report transparent, logical, and readable.
6. Discussion and presentation		**How well is the general presentation supported by the methodological and substantial discussion?**

If the presentation gives reliability answers to two decimal points but fails to have the same answer in different places, this adds uncertainty. The scientific way is to present the results with uncertainty, preferably quantified.

5.2.7. Summary of Conducting a Simple Study

This section has suggested a simple methodology and a design for a study to collect, analyze, and present reliability data based on a review of the literature and the scientific principles of such a study. This has been concretized with a short review of the steps based on Norwegian conditions, rules/standards, and choices. Other choices give other answers, and other areas need attention, but it must not be forgotten that a study in a larger geographical area (state or nation) must use proven scientific methods. Without a correct approach *together* with the discipline of fire science, this will only cast doubt on the study.

A more complex study, an explanatory (causal) study with a mix of intensive and extensive design and many units and many variables, is explained in the next section.

5.3. How to Perform a Complex Study (Explanatory Study)

Based on the overview in Table 5.1., in this section the authors make suggestions for a complex explanatory study. The steps in a descriptive study are also a part of an explanatory study. The steps are not elaborated to any great extent; the goal is simply to convey the meaning and give the general idea for a complex study.

5.3.1. Step One: Development of Problem and Purpose

The first step is to create an overall design and define what is of interest in the study (Table 5.6.). This includes a detailed plan of what kinds of sprinkler systems are of interest, how, where, and over what time, as well as reasons for sprinkler systems not to function as designed.

TABLE 5.6
Design of complex study

Main step	Sub step	Task
Preparation and collection		
	1.1. Is the issue clear or not?	The purpose of the study is to find the reliability (to function as designed) for sprinkler systems in Norway and the reasons for not functioning as designed. Design means by chosen sprinkler systems standard.
	1.2. Is it descriptive or explanatory (causal)?	This is both a descriptive and explanatory study, because there is a wish to determine the reasons why sprinkler systems sometimes do not work as designed.
	1.3. Is it desirable to generalize or not?	It is desirable to generalize from historical reliability, as this is a good indicator of future probability, if the sprinkler systems are designed and installed under the same conditions as those in the study.

Main step	Sub step	Task
	Preparation and collection	
1. Development of problem and purpose		**Conduct a descriptive and explanatory study that generalizes the national reliability and reasons why Norwegian sprinkler systems do not work as designed.** **The reason for a causal study is to be able to make suggestions for improvement.**
	2.1. Intensive (deep), extensive (width) or both study design	Based on the purpose, a mixed intensive and extensive design with many units and many variables is chosen.
	2.2. Descriptive or explanatory	An explanatory design is chosen; this makes some demands on how the study is conducted. 1. Correlation between cause and presumed effect. 2. Cause must precede effect in time. 3. Control of all other relevant factors.
2. Choice of overall study design		**Create a mix of an intensive and an extensive study with a descriptive overall study design, based on three steps of correlation.**
	3.1. Operationalization, how to make a concept measurable	Definition: *Sprinkler system activation*: 1: the sprinkler control valve (alarm valve) opens; 2: the pump (if installed) starts; and 3: the sprinkler alarm activates. *Fire controlled by sprinkler system*: the fire is contained by the sprinkler system's design (number of activated sprinklers and square metres they cover) or improved protection against injury and life loss is gained for residential systems. Improved protection for residential systems is quantified as the possibility to escape (alone or with help) within 15 minutes of the start of a fire.[1] *Fire not controlled by sprinkler system*: a fire not contained within design area for the chosen sprinkler design; or improved protection against injury and life loss not gained for residential systems. "Not gained" improved protection is quantified as the inability to escape (alone or with help) within 15 minutes of the start of a fire.[1]

(Continued)

TABLE 5.6 *(Continued)*

Main step	Sub step	Task
	Preparation and collection	
		Cause and reason: these are always of interest, but the form should only look at reasons. For example, can a valve be closed because maintenance is done at the time of a fire, or because of a lack of maintenance (it does not work)? To make the form as manageable as possible, interviews/investigations will follow reports of faults to find the causes.
	3.2. Design of the study	Based on design and event tree analysis, a form is created.
	3.3. Source and use of sources	It is desirable to use the Norwegian BRIS (Brann, Redning, Innrapportering, Statistikk/Fire, Rescue, Reporting, Statistics) for reporting fires.
	3.4. Selection and limitation	Only buildings fully protected by sprinklers or parts of buildings that are separated by fire section walls are of interest. All systems (NS-EN 12845:2004, CEA 4001, NFPA 13, and INSTA 900–1 and 2) will be examined in this study. Data of system type will always be validated against the ESS (Elektronisk System for Sprinkleranlegg/Electronic System for Sprinklers. Fires from 2015–2020 will be examined.
3. How to collect data?		**The detailed design of the study is based on Norwegian BRIS, looking at building fires from 2015–2020 in buildings protected following NS-EN 12845:2004, CEA 4001, NFPA 13, and INSTA 900–1 and 2 rules. Details on sprinkler systems and hazard classes are validated against ESS.**

[1] Since there is no known study or report on the time of the fire start in residential buildings with sprinklers, or the time of improved protection against injury and life loss (though this is quantified), and both the time frame and measurement of injury cease or are taken over by external organizational measures such as the fire service's efforts, the study's goals should likely be adjusted at some point in the process. For example, if the collected data indicate that 80% of the people in the buildings following Norwegian regulations escaped within 15 minutes, but 99% escaped within 20 minutes when the sprinkler system operated as designed, the design of the study should be adjusted. The overall fire design of a building with a sprinkler system that operates as designed has fulfilled its function in 99% of the cases within 20 minutes, and the authorities can use these data. There is also the possibility of controlling a fire if there is development over time, and the authorities can adjust or accept the demand for fire safety. It also worth considering whether improved protection residential systems should be quantified. If 20 minutes is the time needed for 99%, does this have implications for the design of such sprinkler systems?

With a specific overall design in place, an event tree analysis can be performed.

5.3.2. Step Two: Choice of Overall Study Design

In the event tree shown in Figure 5.4., the starting point is: Fire in sprinkled building. Even if data collectors are not totally sure what each outcome will bring, this can be handled in a review (analysis in Table 5.7.) of each outcome shown in the right side of the figure.

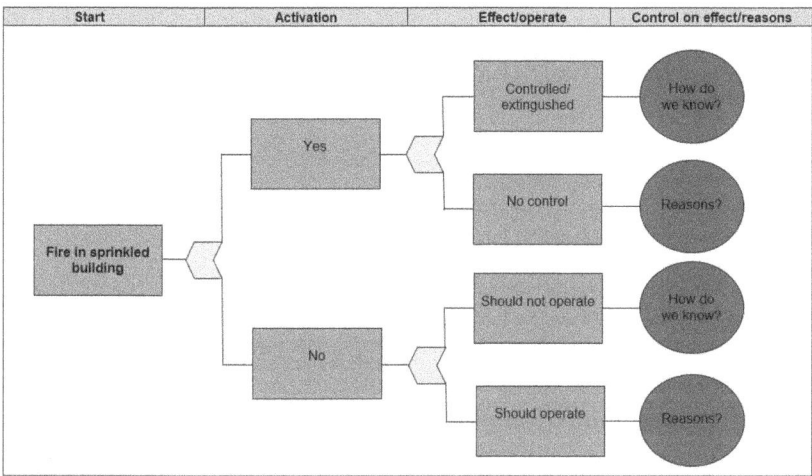

FIGURE 5.4
Event tree, complex design.

TABLE 5.7
Review of reasons for fire not controlled/not extinguished when system has operated

Nr.	Reason	Comments	In form or not?
1.	Inspection and maintenance.	Lack of inspection and maintenance (both that done by the owner and external) is perhaps the most common fault. This is the probably the most under-reported area in most studies today, because it is difficult to separate from faults/damage. When an inspection and maintenance program is followed, most faults and damage on sprinkler heads, pipes, valves, closed or semi-closed valves, alarms, and pumps can be discovered and fixed. So, where is the line drawn between component faults and maintenance? Why are data collected on component faults, if they are finally seen as a failure of maintenance? These are difficult questions, but for this study, both data about component faults and data about maintenance will be collected and analyzed.	A question already asks about maintenance, so further questions will not be asked in this study.

(Continued)

TABLE 5.7 *(Continued)*

Nr.	Reason	Comments	In form or not?
2.	Damage to sprinkler heads.	This is a common fault of sprinkler systems discovered in inspections, especially in buildings with some sort of storage. A sprinkler head that is damaged will have incomplete and/or disturbed water acquisition; this influences the process of getting control of or extinguishing the fire. One challenge is that on a standard system, one or perhaps a few damaged sprinklers should have little influence on controlling the fire within the design area. The reason for having different design areas is to provide a robust design that can handle small faults. A second challenge is that if investigators only look for damaged heads where there was no fire control, they may reach the wrong conclusion. If this is to have scientific value, the information should be collected for all the systems.	In most cases, it will be possible to learn this from rescue personnel on site. This should therefore be included in the form.
3.	System components damaged.	In addition to sprinkler heads, damage could include pipes, valves, or pumps. Since this is often less obvious, damage will be included as a question.	Indicative question, where different components are listed (need follow-up).
4.	Obstacles to water distribution.	Another common fault is an obstacle so that water does not reach the fire. This can be an obstacle to the spray pattern in the ceiling (beams, fixtures, signs, and so on) or a shield over the flammable material. Shelves and covers of different types are the most common. While obstacles in ceilings have the challenges discussed above, obstacles over the flammable material have different challenges. Obstacles in the ceiling prevent some water from reaching the fire, but new sprinklers activate and surround the fire (at least in theory). However, a dense shelf in a rack will not only shield the fire from water, but will also contribute to the fire growth (reflecting the heat and leading the fire plume out to the sides, preheating the material stored beside the fire). This can overcome the sprinkler system (depending on the affected area). If the dense shelf is made of flammable material (e.g. chipboard), there is reason to assume that it will disappear to a greater or lesser extent.	This will be divided into two questions: Are there obstacles in the ceiling that influence water distribution to the sprinkler? Are there reasons to think that there have been obstacles, like a dense shelf, over the area with flammable material? Both questions can be included in the form; the second needs follow-up.

Nr.	Reason	Comments	In form or not?
5.	Inappropriate system for type of fire.	An inappropriate system can be a sprinkler system, where the storage method is not in consensus with the sprinkler standards, e.g. tire storage, and another system should have been chosen, e.g. a foam system. Alternatively, there may have been a change of use or storage, and the hazard class has gone up, without the design criterion for the sprinkler system being changed. Even if there are indicative clues for rescue personnel, there is no reason to demand that they should notice them or understand all parameters that must be considered to determine the correct system for the type of fire. Even if data on this are interesting, they have nothing to do with unreliability. Sprinkler reliability is the ability to function as designed. If the system is designed incorrectly or is not redesigned after changes in use/storage, this has nothing to do with the ability of the system.	Indicative question used if there are reasons to think an inappropriate system was present at the time of fire.
6.	Not enough water discharged.	Three situations can result in not enough water discharging. First, the sprinkler system is inappropriate for the present hazard class (designed for a lower hazard class with lower water density). Second, there is not enough water for the design. This is something that should be noted at the time of installing/commissioning or inspection; maintenance should have revealed it, if changes to available water occurred after commissioning. Third, there is not enough water for the sprinkler and the fire fighters. Norwegian regulations do not demand water for both.	This is covered in point 5 above.
7.	Manual intervention.	Manual intervention, like closing valves, has a negative effect if it is done at the wrong time. This does not have anything to do with a properly designed system, but is a "fault" that lies outside the design area.	This is covered in point 5 above.

(Continued)

TABLE 5.7 *(Continued)*

Nr.	Reason	Comments	In form or not?
8.	Not installed as designed	What if the sprinkler system is not engineered and/or installed as designed? What if the main valve is not supervised as demanded in the design? What if the hydraulic calculations have not included sprinkler fittings (flexible hose), and so on?[1]	Include in the form, with follow-up.
9.	Other.	Data collectors must keep an open mind and acknowledge that science is the pursuit to understand what is yet not known. Without being able to list what is not yet thought about, progress is not easy.	Include in the form, with follow-up.

[1] Should a system be rated based on "wrong" design, or should some robustness be assumed? Perhaps some of each? Perhaps some functions should not be treated as if the system has been in operation over one year and faults should have been discovered during inspection/maintenance, for example, if a valve is not supervised and is closed outside working hours. The big challenge is that there is little knowledge of this, since there is no platform to look at these things.

Outcome 1 (Controlled/extinguished: How do we know?): If we want to make sure that the person filling out the form has understood the outcome as success (controlled/extinguished) right, both the number of heads and the fire affected areas must be collected.

Outcome 2 (No control: Reasons?): There are several reasons why a sprinkler system is unable to control a fire (or extinguish a fire, as for example, ESFR systems), either as sole reasons or in combination. The list in Table 5.7. gives various options.

Summary: "Damage to sprinkler heads" and "Obstacles to water distribution" will be incorporated into the form so that the person filling in the form must answer them for both controlled/extinguished and not-controlled outcomes. There will be an indicative question about "System components damaged (pipe, valve, or pump)" and "Inappropriate system for type of fire (wrong type or wrong hazard class)." Collecting data on inspection and maintenance will be very important.

Outcome 3 (Should not operate: How do we know?): See Section 5.2.2, "Step Two: Choice of Overall Study Design."

Outcome 4 (Should operate: Reasons?): If there is no understanding of the reason for a fault, whether it is a sprinkler fault or a human/technical fault outside of what a sprinkler system is designed for, the outcome will be placed in the wrong category. It is not within the design of a sprinkler system to overcome manual intervention, for example, closing the main valve, and, therefore, this has nothing to do with unreliability.

Table 5.8. reviews the reasons a sprinkler system does not activate and operate but should.

TABLE 5.8
Review of reasons for system not activated/operated when it should have operated

Nr.	Reason	Comments	In form or not?
1.	Inspection and maintenance.	See footnote 1 to the Table 5.7 under Outcome 1 above.	Already included.
2.	Sprinkler head.	Both the Omega failure and counterfeit sprinklers clearly show that this can affect both activating and operating. Furthermore, there is little knowledge about how sprinklers over a given age (if any) react to temperature and if this affects the RTI-factor.	Include in form, even if there are not enough resources to inspect affected sprinklers in test laboratory.
3.	Pipe.	If a sprinkler activates, but there is no water or visible low flow of water, there are reasons to examine the state of the pipe. Rust, MIC,[3] or physical damage to pipes could explain why sprinklers do not function as designed.	Include in the form, with follow-up.
4.	Valve/pump.	Valves may fail for several reasons: lack of maintenance; closed; no supervision/ monitoring; sabotage. Without collecting and separating reasons that should and should not be included in unreliability and separating the types of valves (dry, wet, pre-action/deluge), it is not possible to have built-in reliability/ unreliability. The same could be said for pumps, but there is also the question about power supply or the use of diesel pumps.	Include in the form, with follow-up. Collection of data on types of pumps must also be done.
5.	Other.	Without being able to list what has not yet been thought about, progress is not easy.	Include in the form, with follow-up.

[3] Microbiologically influenced corrosion is a form of localized corrosion. Material is lost at discrete points, instead of universally across an entire surface.

Summary: All five reasons for not activated/operated will be incorporated as indicative questions with follow-up.

5.3.3. Step Three: How to Collect Data

Step three is to write the form. This is basic, regardless of how the form is distributed or how information is collected from databases. The main purpose is to have control of questions that are of interest and to use them in the quality assurance of the analysis. Table 5.9. gives a sample questionnaire.

As pointed out in Section 5.2.3. on collecting data, there has be some sort of cooperation with fire brigades. Even if it can be expected that a system

TABLE 5.9

Complex inquiry form for fires in buildings protected by sprinkler systems

Form for fires in buildings protected by sprinkler systems	

1. Address: — Official identification number (Norway Gnr/Bnr)

   Type of building: — Is the building registered in ESS? If not, must the owner do so?

   Date and time of fire:

   Installing year and latest inspection/ maintenance: — Data from documentation or from ESS.

Information about the sprinkler system

2. Type of sprinkler system and type of main water supply: — CEA 4001
   NS-EN 12854
   NS-INSTA 900
   NFPA 13
   Other (specify)
   Water tank

3. Hazard class: — OH/HHP/HHS
   1–4

Information about consequences of fire

4. Did the fire activate the sprinkler system? — Yes (point 5/6)/No (point 9)

5. If yes, did the sprinkler system activate the alarm and/or the pump? — Alarm (internal or external)
   Pump (if present)

6. If yes, did the sprinkler system control/ extinguish the fire? — Yes (point 7/8)/No

7. If yes, how many sprinklers were activated, what was the area of flame damage, and was there any damage to the sprinkler head? — Number:
   m^2:
   Damaged sprinkler:

8. If yes, are there any indications of obstacles in the ceiling and/or over the flammable material? — Ceiling:
   _____.
   Stored material: _____

9. If no, what is the area of flame damage? — m^2 (point 10):

10. If no, are there any indications that the system should have operated? — Yes (point 11/12/13)/No (point 14/15/16)

11. If yes, are there any indications of faults/ damage to the sprinkler, pipe, valve, pump, or anything else? — Sprinkler: Yes (Number)/No
    Pipe: Yes (_____)/No
    Valve: Yes (_____) /No
    Pump (if present): Yes (_____)/No
    Other: _____

12. If yes, are there any obstacles in the ceiling and/or over the flammable material? — Yes (_____)/No

13. If yes, does the alarm test of the sprinkler system work? — Yes/No

14.	If no, are there any indications or faults/ damage to the sprinkler, pipe, valve, pump, or anything else?	Sprinkler: Yes (Number)/No
		Pipe: Yes (_____)/No
		Valve: Yes (_____) /No
		Pump (if present): Yes (_____)/No
		Other: _____
15.	If no, are there any obstacles in the ceiling and/or over the flammable material?	Yes (_____)/No
16.	If no, does the alarm test of the sprinkler system work?	Yes/No

* (____) Written comments.

like BRIS collects needed information on sprinkler activating (like the form designed in Section 5.2.3.), there are no reasons to expect that fire brigades can collect all this explanatory information. This most likely must be done by the members of the research team.

5.3.4. Step Four: Analyzing and Calculating, Including Uncertainty

While it may lead to more work in the long run, the researcher's personal involvement gives the possibility of checking the value of the basic information received. This could bring new depth and understanding to the numbers used.

When it comes to the descriptive part of the study, the points given in Section 5.2.4. also apply here.

There is one more possibility that needs to be examined. With a detected fault, what information can be gained by studying former inspection and maintenance reports? What do they tell about the condition of the sprinkler system? Were there deviations in crucial areas like water supply, maintenance of alarm valves and pumps, or testing? For example, should a fault in a sprinkler pump that was detected one year earlier be listed as a pump failure, or as missing maintenance? With a system that demands repair and follow-up, it is presumed that such faults are not so common. This must be considered by the research team in the context of uncertainty and be included in calculations.

Another consideration is how inspections are carried out. In some countries, fire officials do this; in others, volunteers handle it. Both influence results.

Working with such questions and receiving information is part of step five described in the next section; the authors simply want to point out that these processes go together.

5.3.5. Step Five: Quality Assurance of the Analysis

With good quality assurance, the fundamental requirements for a solid presentation in the last step are mostly met. The importance of checking the answers given by the analysis against the purpose of the study cannot be stressed enough. Table 5.10. gives a minimum layout for the form based on the purpose of the sample complex study; it can be extended as needed.

TABLE 5.10
Design of complex study, step five

5.1. Conceptual validity	**Overall question: Do the indicators measure what is of interest?** 1. How many building fires were there per year during 2015–2020? 2. How many fires were in sprinkled buildings? 3. How many sprinkler systems were CEA 4001/NS-EN 12845/NS-INSTA 900/NFPA 13/Other system? 4. How many fires activated the sprinkler system (both overall and by alarms and pumps)? 5. How many fires were reported as controlled/extinguished? 6. How many sprinklers were activated for different systems and hazard classes? This includes those given as non-controlled if they were, in fact, controlled. 7. When the system did not activate, was this linked to the sprinkler bulb/fusible link being destroyed/damaged, or fault/damage to pipe, valve, or pump (if present)? 8. Are there indicators that the system should have activated/worked? 9. Did the alarm test work on the non-activated sprinkler system?
5.2. Validation of correlations[1]	1. How many fires were controlled/extinguished according to number of sprinklers and area of damage? 2. Is the area affected by the fire (flames/hot plume) correlated to the number of activated sprinklers? 3. Is the area and the number of sprinklers activated for a non-controlled fire correlated to the stated effect?[1] 4. What influence do obstacles in the ceiling have? 5. What influence do obstacles over stored material have?
5.3. External validity	1. Are some geographic areas or fire brigades excluded from this study? 2. Do the findings support known theory? 3. Is uncertainty quantified?
5.4. Are the results trustworthy?	Once again it is important to ask: Does the way the data are collected (e.g. the inquiry form) and analyzed produce the result? Three possible problems with the conclusions are level failures, time frame failures, and causality failures.

	An example of a *level failure* is when the results show that most fires are confined to the room of origin, and the conclusion is that sprinklers are designed to confine fires to the room of origin. Do data support this and does the design follow standards and test protocols? When the conclusion takes the result up or down a level, these failures occur. An example of a *time frame failure* is the assumption that a sprinkler system is no less reliable because of its age. For example, a study that looks at a selected sprinkler system from installation to five years later may conclude there is no difference in reliability based on age. An example of a *causality failure*[2] is the lack of a cause of the failure to activate. The conclusion may be that it should not activate, with no data to support this conclusion.
5. Quality assurance of the analysis	**Based on validation of concept, correlations, external validity, level failures, time frame failures, and causal failures, how good are the conclusions drawn from the analysis?**

[1] Uncertainty can be handled theoretically by only looking at the collected data, or more practically by further investigation. Case studies (in fact, these are follow-ups) give more insight into how fires and sprinklers are connected and yield more correct uncertainty data.

[2] The authors conclude causality is based on three conditions: correlation of cause and presumed effect; cause precedes effect in time; and control of all other relevant factors. It is not enough to have the first two.

When conducting the quality assurance (QA) analysis, it often happens that the form must be corrected, and the analysis must be done over again in some areas, but this is to be expected. Quality assurance processes may reveal areas not previously thought about, but the data may be collected and cannot be changed in an economical/practical way. In these cases, comments about uncertainty will strengthen the sense that the study has been conducted in a scientific way.

5.3.6. Step Six: Discussion and Presentation

The last step is to write the report and present the results of the study. The areas shown in Table 5.11. are key.

5.3.7. Summary of Conducting a Complex Study

Following Section 5.2., "How to Perform a Simple Study (Descriptive Study)," this section has suggested six steps for a complex causal study, based on Norwegian conditions and the authors' knowledge of sprinkler systems. This kind of work needs a thoughtful and systematic approach, and this cannot be stressed enough.

TABLE 5.11
Design of complex study, step six

Presentation		
	6.1. Methodological discussion	A theoretical review of the methodology, including study design, data collection, analysis, and quality assurance of the results. 1. How good is the reliability of the data? 2. How good is the conceptual validity? 3. How good is the internal validity? 4. How good is the external validity?
	6.2. Substantial discussion—connection of findings and theory	*Findings*: Are the results consistent or inconsistent with earlier and other similar studies? When there are earlier studies, what are the long-term changes? *Theory*: What is the connection between results and known theory? What is unknown? Could some hypotheses be followed up in later studies? *N.B.*: What do the findings tell about the sprinkler standards that have been used?
	6.3. Presentation (also uncertainty)	There must be a definition list or presentation of key terms earlier in the report. A summary or abstract of findings helps the reader understand the report. Present the results with quantified uncertainty. Make the report transparent, logical, and readable.
6. Discussion and presentation		**How well is the general presentation supported by the methodological and substantial discussion?**

Since this has been an explanatory study, and reliability is defined as the ability to work as designed, the authors have also touched on unreliability/ failure. More information is given in the next chapter.

It is important to make forms manageable for those filling them out. If they find the forms long and difficult, the quality of the answers will probably drop.

The type of study described here will give lasting scientific value.

6

Conclusion

This section of the book sums up some of the issues covered, the kind of data investigated to answer the research questions, and the findings and their importance. It offers some conclusions, gives possible explanations of the findings, notes their implications, acknowledges the limitations of the work, and make suggestions for further work.

6.1. Summary

The authors started writing this book with a desire to learn the reasons for the diversity in the key terms and levels of reliability from one report to another, both within countries and between countries. After reviewing the literature, they came up with a list of five publications that they wanted to examine more closely: three by NFPA (1970, 2010, 2017), one by Marryatt (Marryat, Rev. 1988), and one by NFSM (Optimal Economics, 2017).

In the "Automatic Sprinkler Performance Tables, 1970 Edition" (National Fire Protection Association, 1970), the original data from 1897 to 1924 are updated to include data from 1925 to 1969. Key definitions like control—the "prevention of excessive fire spread in light of the nature of the occupancy"—are illustrated in a graph showing the differences in the number of sprinklers operated in wet and dry systems. The fact that the report does not use the lowest performance rate raises some questions.

In the 2010 report from NFPA, "U.S. Experience with Sprinkler and Other Automatic Fire Extinguishing Equipment" (National Fire Protection Association Research, 2010), there are major changes in the area of interest (all extinguishing systems) and the methodology. The use of NFIRS 5.0 as a prime data source leads to the report's conclusion that 49% of all fires were too small to activate the systems. The report also states that effective performance is indicated by confinement of fire to the room of origin, and effectiveness declines when more sprinklers operate. None of these conclusions is proven or substantiated.

In the more recent "U.S. Experience with Sprinklers" (National Fire Protection Association Research, 2017b), the focus is again on sprinklers, and home fires are given special attention. All data are from NIRS 5.0. The 2017 NFPA report focuses on five or fewer heads operated when the sprinkler design for such buildings is two, or up to four, but there is no explanation of the reason for this. As in the two previous reports, no long-term trends are included.

The comprehensive book by Marryatt (Marryat, Rev. 1988), *Fire—A Century of Automatic Sprinkler Protection in Australia and New Zealand—1886–1986*, gives the following explanation of high reliability: "Inspection, testing, and maintenance exceeded normal expectations, and higher pressures." However, all the systems in this study are wet systems; it excludes all fires where sprinklers were shut off. The fact that most of the cases come from Wormald International Group of Companies and the fact that 99.5% reliability refers to up to 113 sprinklers are probable explanations to the high reliability.

The last publication is from the UK, "Efficiency and Effectiveness of Sprinkler Systems in the United Kingdom: An Analysis from Fire Service Data" (Optimal Economics, 2017). There are calculation errors, and there are no explanations of the inconsistent numbers. Different numbers seem to be used in each case. However, this is the only report to use the word "indicates" about its findings.

The review triggered the quest to understand how these reliability data were collected, analyzed, and presented. The authors needed a systematic tool to validate the studies. They opted for document analysis, basing the analysis on how a scientific investigation should be carried out. The discovery was that all the publications had problems in four out of eight possible areas.

Document analysis is primarily a tool for social science, but as this book shows, it is very useful in the field of fire science. Therefore, the authors have proposed a modified methodology adapted to the scientific principles of fire science. They have made two proposals for future study using the suggested methodology, the first descriptive and the second explanatory.

6.2. Conclusion

The book begins and ends with a caution: there is a lack of knowledge of what an extinguishing system is (does it extinguish or control a fire, or both?). Moreover, there are different types of sprinkler systems, including several different types of residential systems, systems for ordinary hazards, storage, and special systems, but there is little trace of this in literature. Finally, different sprinkler systems perform differently, and this is not acknowledged. A recognized comprehensive book on fire science has much more information on hydraulic calculations than on extinguishing theory or the use of different systems.

The same can be said about reliability. The perception that all sprinkler systems should be treated equally in relation to reliability is wrong. None of the publications has a definition of reliability or supports its views by referring to the science of reliability. This is a major finding. None of the publications examined in these pages communicates or works with the fact that reliability is the system's ability to function as designed to a specific sprinkler standard.

More surprising is the "circle of trust." Even if there is a reference to sources, this does not mean it exists or is available. It is not certain that the authors have the correct result. It is not even certain that they have even read the reference. Because some studies are comparative, there is an appearance of much data, but this is not the case.

In the validation of the studies using document analysis, it became clear that they all had problems in four areas: unclear issues, including missing definitions and intentions of the investigations; uncertain data collection processes; varying quality of analysis and lack of quality assurance; and lack of systematic presentation and discussion. Based on this finding, the authors conclude that none of the reports on sprinkler reliability can be used as a general documentation of reliability or of the future probability for sprinkler systems to function as designed.

Are these findings and conclusions supported by other scientific evidence? The overall evidence is that these findings are correct. First, the list of the literature is much shorter when the references that are impossible to find, comparative studies that use incorrect findings, studies with a small- or limited-time frame, and older studies are removed.

Second, the critical review and the analysis of the selected publications indicate most do not follow recognized scientific principles.

Third, both government publications (Department of Building and Housing, 2005) and other publications (Frank, Gravestock, Spearpoint, & Fleischmann, 2013) show a lack of information. Their conclusions and methods are not clear.

Why are these findings important? First, there is little knowledge about sprinklers and their effect on live fires outside a test facility, the kind of protection they give, or the time before a flashover in different types of sprinkled buildings can occur. Second, the criteria for selecting and validating the "right" type of sprinkler system do not seem to be established. Third, there is no known reliability for each system or hazard class.

As this book makes clear, there are no data on the reliability of the different types of sprinkler systems. There is no knowledge of reliability measured in terms of the operationality and efficiency required by the various sprinkler standards. We do not know if today's division into hazard classes, design area, and water density is adequate. Finally, standards are not revised using data on how sprinklers actually behave under fire. Revisions come from studies of single cases and tests done in laboratories.

The book makes a major contribution to the field by developing a new methodology to validate scientific studies. There is now a possibility of validating work systematically.

Sprinklers have been around for 133 years (1886 to 2019), so why is there not more knowledge? This is a good question, and the authors have some, but not all, of the answers. The first answer is probably the most surprising one: sprinklers work. Most of the authors' experience with this kind of system suggests they do work. The problem is that the overall reliability of all sprinkler systems over their lifetimes is not known; the reliability of each type is not known, and the effect of age is unknown.

Unfortunately, changes in methodology have been small. Without a systematic and critical look at methodology, which is the basis of science, improvement is difficult. The fact that large and important organizations can conduct their research without outside influence and have their own interest in publishing good results stresses the importance of independent research.

The second answer is the lack of teamwork. Good scientific work in collecting, analyzing, and presenting reliability demands that the people involved have a broad understanding of and detailed knowledge about fire science, sprinklers, how to conduct studies, how to understand the collected data, how to analyze the data, including uncertainty calculations, and how to present the findings. In addition, there is a need to cooperate with the proper authorities to get the right data or adapt data to the research.

Finally, decision makers seem to know little about the area. They question the use of time and money to improve an area they erroneously think they know. The real question is, will this be done now that the errors have been revealed?

6.3. Suggestions for Further Work

The Master's thesis and this book have the following implications:

1. The fact that 29% of the references were either incorrect or non-existent indicates the need for basic research conducted in a proper way and, sadly, for readers to avoid assumptions that presented data are correct.

2. The fact that there are no reliable data on sprinkler reliability means the fire community does not have access to the accuracy needed in fire safety engineering.

3. Even if the authors have not examined the reliability data for other active and passive fire protection measures, is it clear that that no

other measures have adequate data and studies. Available data strongly suggest that the situation is not better for them; in fact, it is likely worse.

4. Performance-Based Fire Protection Design says: "In the analysis of an existing building, the type (smoke detection, heat detection, UV/IR) of an automatic detection system must be documented. . . . Similarly, whether in an analysis of an existing building or in the design of a new building, the characteristics of automatic suppression systems must be documented" (Society of Fire Protection Engineers, 2016: p. 1 265). The design's close links to quantitative and qualitative data have several implications:

Comparative criteria: Since there is no knowledge of the performance of sprinkler systems either in general or for specific types, how should comparisons be made?

Deterministic criteria: How can it be shown that the worst case scenario will not happen, when it is not possible to prove that, for example, the sprinkler will not work in more than 50% of the cases?

Probabilistic criteria: How is it possible to set the probability of a given event acceptably low, when it is now clear that data are limited and unreliable?

There are, of course, many limitations to this book. This is a subject that could be investigated in much greater depth, but the basic research had a time limit. Many articles and studies have key words like sprinkler and reliability, even if they only briefly touch on the theme, making it hard to track the right publications. Some can be missed just because of the sheer number of hits in an online search. It was also surprisingly difficult to get data and literature from some major world contributors within the fire community; in fact, the main author of this book did not receive any information in some cases. The fact that some authorities stopped communicating after being asked how data are obtained illustrates the difficulty of starting a good dialogue.

Suggestions for further work:

1. There is an urgent need to start all over again with basic research on fire protection measures (active and passive) and their reliability.
2. There is a need to develop fire and extinguishing theory, based on proper research, both under laboratory conditions and for real fires.
3. In anticipation of new reliability data collected using the principles shown in this book, methods must be developed to use the data in such way that they give scientific value.
4. The methodology presented in this book should be expanded to include other active and passive fire protection measures.

6.4. How Scientific Is the Use of Document Analysis?

The use of document analysis is a well-established scientific method for the systematic analysis of documents, but two aspects require further attention.

First, how well does this method from the social sciences work when it is applied in the natural sciences, in this case, Fire Safety and Fire Engineering? In Chapter 5, the authors give suggestions on methodology and designs for scientific investigation/studies, with a focus on collecting, analyzing, and presenting data on sprinkler reliability. These can, of course, be used in other areas as well. Adaptation must be transparent, so that if others are doing the same type of study, they will come to the same result.

Second, should all questions be answered "Yes" in the document analysis to make a study a scientific one, or is it possible to score some "No" or "Not sure" answers without being "unscientific"? Mainly, this depends on the purpose, sources, and resources available. For example, consider the three first questions under "development of problem and purpose":

1. Is the issue clear?
2. Is it explanatory (causal) or descriptive (descriptive)?
3. Can it be generalized?

First, if the purpose or issue of the study is not clear, this often means it suffers from a lack of definitions. Alternatively, the purpose may not have been followed or adjusted to suit available sources and data. Lack of sources or data can be compensated for and overcome. If a fire brigade report lacks central information, for example, a written inquiry, telephone contact, or even a personal visit/investigation is possible, but this comes down to what kind of the resources the study team has. If there are not enough resources/time available, the purpose should be adjusted to reflect this.

Second, a study can be both explanatory (causal) and descriptive. None of the studies examined in the literature review received a "Not passed" for the first question, "Is the issue clear?", because it was either explanatory or descriptive. The challenge is that the amount of data increases exponentially when a study changes from descriptive to explanatory. Are the researchers more interested in how (descriptive) or why (explanatory)? As soon as the interest shifts from whether something works or not to what makes it work or not, the purpose, data, and resources must be adjusted. This is often not the case, and the causes of interest are not redefined. Interestingly, few (perhaps no) studies look at why sprinklers work in different buildings, with different storage configurations and human behaviour.

Finally, if a study wants to define general reliability, the demand to generalize often pushes the benchmark up, not down. The most common fault in the studies examined is the assumption that all sprinkler systems have the same function. This is not the case; an NFPA 13D system cannot be compared to an ESFR system. This may have been a conscious generalization, or it may reflect a lack of study management. In either case, it is not possible to make conclusions about the reasons for this, from this study.

Glossary

The definitions are based on *Kollegiet for brannfaglig terminologi* (Kollegiet for brannfaglig terminologi, u.d.), if not specified otherwise.

Active fire protection measures: Technical fire protection with a function activated after fire is detected, auto fire alarm is triggered, or officials are notified.

AES: Short form for Automatic Extinguishing Systems (National Fire Protection Association Research, 2017).

Data analysis: The process of developing answers to questions through the examination and interpretation of data. The basic steps in the analytic process consist of identifying issues, determining the availability of suitable data, deciding which methods are appropriate for answering the questions of interest, applying the methods, and evaluating, summarizing, and communicating the results (Statistics Canada, 2018).

Engineer: Person with technical education (Store norske leksikon, 2018).

Engineering: Engineering consists of making calculations, overviews, drawings (principle and detail), and descriptions of a project (Store norske leksikon, 2018).

Fire alarm system: A system for fire detection and alarm with fire detector, alarm device, central control unit, and, if necessary, an orientation board.

Fire extinguishing equipment: Manual or automatic system designed to extinguish or control a fire.

Flashover: Transition to a state where all surfaces of combustible materials in a room participate in a fire.

Ignition: The start of combustion.

Inspection and maintenance: Inspection is the examination of status in relation to requirements, and maintenance refers to repairs, replacements, remedies (defects, errors, and omissions), testing, and servicing of active and passive fire protection measures so they function as required.

Operational reliability or operationality (not to be confused with reliability) is a measure of the probability that a protection system or part of it will operate when needed.

Performance reliability or efficiency (performance is probably not the best word; efficiency is likely better) is a measure of the adequacy of the system to successfully perform its intended function under specific fire scenario conditions (fire hazard).

Reliability: The ability to function as designed. Sprinkler reliability is the ability to function as designed to a sprinkler standard and a defined fire hazard (Barry, 2002).

Residential sprinkler system: Simplified sprinkler system adapted to residential housing.

RTI: RTI is a short form for Response Time Index measured in metres/second$^{1/2}$. A fast response sprinkler has a thermal element with an RTI of 50 $(Ms)^{1/2}$ or less; a standard response sprinkler has a thermal element with an RTI of 80 $(Ms)^{1/2}$ or more (National Fire Protection Association, 2016).

Sprinkler head: Nozzle for spreading water as part of a sprinkler system. The nozzle may be open or equipped with a thermally sensitive opening mechanism.

Sprinkler or fire pump: Pump systems, including automatic starting devices, used to obtain sufficient water or apply sufficient pressure to the sprinkler system.

Sprinkler system: Automatic stationary system that aids in the detection and control of fire to prevent flashovers in the room of fire origin, to improve the chances for occupants to escape or be evacuated, and to control or extinguish a fire. A sprinkler system consists of system valves, piping, and sprinkler heads, with water as the primary extinguishing agent.

Water density: Number of litres of water per square metre and minute.

References

Aven, T. (2006). *Pålitelighets og risikoanalyse* (4. utgave. utg.). Oslo: Universitetsforlaget.

Barry, T. F. (2002). *Risk-Informed, Performed-Based, Industrial Fire Protection* (1. utg.). Knoxville, TN: Tennessee Valley Publishing.

Bodil Aamnes Mostue og Kristen Opstad ved SINTEF. (2002). *Effekt av brannverntiltak— Vegger og sprinkler.* Trondheim: Norges Branntekniske Laboratorium AS.

Budnick, E. K. (2001). "Automatic Sprinkler System Reliability." *Fire Protection Engineering*, (9) (ISSN 1524–900X).

Bukowski. R. W., P. E. (1999). Estimates of the Operational Reliability of Fire Protection Systems. *International Conference on Fire Research and Engineering, Third. Proceedings. SFPE and NIST and IAFSS* (ss. 87–98). Chicago, IL: Society of fire Protection Engineers.

DCLG. (2012). *IRS Help and Guidance.* London: Department for Communities and Local Government.

Department of Building and Housing. (2005). *Determination 2005/109: Single Means of Escape from a High-rise Apartment Building.* Wellington, NZ: New Zealand Department of Building and Housing.

Drysdal, D. (1998). *An Introduction to Fire Dynamics* (2nd ed. utg.). Chichester: John Wiley & Sons, Ltd.

Fedøy, A. (2018). *Collecting, Analysing, and Presenting Reliability Data for Automatic Sprinkler Systems.* Western Norway University of Applied Sciences, Department of Fire Safety and HSE Engineering. Haugesund: Høgskulen på Vestlandet. Hentet fra https://hvl.no/bibliotek/

Finucane, M. a. (1987). Reliability of Fire Protection and Detection Systems. *Recent Developments in Fire Detection and Suppression systems* (s. 20). Edinburgh, Scotland: University of Edinburgh, Unit of Fire Safety Engineering.

Fire engineering. (1997, September 1). *www.fireengineering.com.* Hentet fra silent-sentinels-under-fire: www.fireengineering.com/articles/print/volume-150/issue-9/departments/editors-opinion/silent-sentinels-under-fire.html

Frank, K., Gravestock, N., Spearpoint, M., & Fleischmann, C. (2013). *A review of sprinkler system effectiveness studies.* Fire Science Reviews. Hentet fra https://doi.org/10.1186/2193-0414-2-6

International Maritime Organization. (2008, May 9). MSC.265(84) Amendments to the revised guidelines for approval of sprinkler system equivalent to that reffered to in SOLAS regulation II-2/12 (A.800(19)). London, United Kingdom. Hentet fra www.imo.org/en/KnowledgeCentre/IndexofIMOResolutions/Maritime-Safety-Committee-(MSC)/Documents/MSC.265(84).pdf

Jacobsen, D. I. (2015). *Hvordan gjennomføre undersøkelser* (3. utgave. utg.). Oslo: Cappelen Damm akademisk.

Kim, W. K. (1990, November). "Exterior Fire Propagation in a High-Rise Building," a Master's Thesis. Worcester, MA: Worcester Polytechnic Institute.

Koffel, W. E. (2006). *Final Statement of Reasons for Proposed Building Standards of the Office of the State Fire Marshal Regarding the Adoption by Reference of the 2006 Edition of the Internatiobal Building Code (IBC) with Amendments into the 2007 California Building Code.* Sacramento: Office of the State Fire Marshal.

Kollegiet for brannfaglig terminologi. (u.d.). *www.kbt.no/*. Hentet 2017 fra www.kbt. no/faguttrykk.asp.

Linder, K. W. (1993). Field Probalility of Fire Detection Systems, Balanced Design Concepts Workshop. Gaithersburg, MD: National Institute of Standards and Technology Interagency Report.

Malm, D. a.-I. (2008). *Reliability of Automatic Sprinkler Systems—an Analysis of Available Statistics*. Lund University, Sweden, Department of Fire Safety Engineering and Systems Safety. Stockholm: Lund University, Sweden.

Marryatt, H. W. (Rev. 1988). *Fire—A Century of Automatic Sprinkler Protection in Australia and New Zealand—1886–1986*. North Melbourne, Victoria: Australian Fire Protection Association.

Maybee, W. W. (1988). *Summary of Fire Protection Programs in the U.S. Department of Energy—Calendar Year 1987*. Fredderick, MD: U.S. Department of Energy.

Miller, M. J. (1973). *The Reliability of Fire Protection Systems*. Norwood, MA: Factory Mutual Research Corporation.

Milne, W. D. (1959). "Automatic Sprinkler Protection Record." I W. D. Milne, *Factors in Special Fire Risk Analysis*, Chapter 9, pp. 73–89. Philadelphia, PA: Chilton Company.

National Fire Protection Association. (1995). *nfpa.org*. Hentet fra nfpa.org/about-nfpa/nfpa-overview/history-of-nfpa: www.nfpa.org/about-nfpa/nfpa-overview/history-of-nfpa

National Fire Protection Association. (1970, July). Automatic Sprinkler Performance Tables 1970 Edition. *Fire Journal, 64*(4), 5 (35–39).

National Fire Protection Association. (2005). *U.S. Experience with Sprinklers and other Fire Extinguishing Equipment 1980–1998*. Quiency, MA: NFPA.

National Fire Protection Association. (2010a). *NFPA 13D Installation of Sprinkler Systems in One- andTwo-Family Dwellings and Manufactured Homes*. Quincy, MA: NFPA.

National Fire Protection Association. (2010b). *NFPA 13R Installation of Sprinkler Systems in Residential Occupancies up to and Including Four Stories in Height*. Quincy, MA: NFPA.

National Fire Protection Association. (2016). *NFPA 13, Standard for the Installation of Sprinkler Systems*. NFPA.

National Fire Protection Association. (2018, April 11). Email from Marty Ahrens. Quincy, MA, USA.

National Fire Protection Association. (u.d.). *nfpa.org*. Hentet October 2017 fra Fire Alarm System Research, Where it's been and where it's going: www.nfpa.org/~/media/files/news-and-research/proceedings/firealarmsystemresearchwmoorekeynote.pdf?la=en.

National Fire Protection Association Research. (2010). *U.S. Experience with Sprinkler and Other Automatic Fire Extinguishing Equipment 2004–08*. Quincy, MA: NFPA.

National Fire Protection Association Research. (2017a). *Home Structure Fires*. Quincy, MA: NFPA.

National Fire Protection Association Research. (2017b). *U.S: Experience with Sprinklers 2010–14*. Quincy, MA: NFPA.

NFPA. (2017). Preventing Warehouse Total Loss Caused By Excessive Ventilation. *SUPDEPT 2017* (s. 6). College Park, MD: NFPA.

NFPA. (2018a, April). *National Fire Protection Association*. Hentet fra List-of-Codes-and-Standards: www.nfpa.org/Codes-and-Standards/All-Codes-and-Standards/List-of-Codes-and-Standards

NFPA. (2018b, April). *NFPA Journal*. Hentet fra News-and-Research/Publications: www.nfpa.org/News-and-Research/Publications/NFPA-Journal/2016/ November-December-2016/Features/Sprinkler-Systems

Opplysningskontoret for automatiske slokkeanlegg. (2003). *Hvordan er kvaliteten på sprinkleranlegg i Norge?* Oslo: Opplysningskontoret for sprinkleranlegg.

Opstad, K. o. (2002). *Effekt av brannverntiltak—Vegger og sprinkler*. SINTEF. Trondheim: Norges Branntekniske Laboratorium AS.

Optimal Economics. (2017). *Efficiency and Effectiveness of Sprinkler Systems in the United Kingdom: An Analysis from Fire Service Data*. Chief Fire Officers Association, National Fire Sprinkler Network.

Powers, R. W. (1979). *Sprinkler Experience in High-Rise Buildings (1969–79)*. Boston, MA: Society of Fire Protection Engineers.

Ramachandran, G. (1998). *The Economics of Fire Protection*. New York: E & FN Spon.

Richardson, J. K. (1985). *The Reliability of Automatic Sprinkler Systems*. Ottawa, Canada: National Research Council Canada.

Smith, F. (1982). "How Successful are Sprinklers." *I Fire Prevention, 82*, 28–34. (s. 7). Fire Protection Association 1982.

Society of Fire Protection Engineers. (2016). *SFPE Handbook of Fire Protection Engineering* (5th ed. utg.). New York: Springer-Verlag.

Statistics Canada. (2018, January 16). *Data analysis and presentation*. Hentet fra www.stat can.gc.ca: www.statcan.gc.ca/pub/12-539-x/2009001/analysis-analyse-eng.htm

Store norske leksikon. (2018, April). *Store norske leksikon*. Hentet fra https://snl.no/: https://snl.no/brann

Taylor, K. T. (1990). "Office Building Fires . . . A Case for Automatic Fire Protection." *Fire Journal*, 52–54.

TYCO. (2005). "Fire Sprinkler History - NFSA, NFPA & Tyco." *The Station House*, 4(1), 8.

Index